IGCSE

Mathematics
Core and Extended

Practice Book

Ric Pimentel
Terry Wall

HODDER
EDUCATION
AN HACHETTE UK COMPANY

® IGCSE is the registered trademark of Cambridge International Examinations.

Answers can be found at www.hoddereducation.com/cambridgeextras

This text has not been through the Cambridge endorsement process.

Hachette UK's policy is to use papers that are natural, renewable and recyclable products and made from wood grown in sustainable forests. The logging and manufacturing processes are expected to conform to the environmental regulations of the country of origin.

Orders: please contact Bookpoint Ltd, 130 Milton Park, Abingdon, Oxon OX14 4SB. Telephone: (44) 01235 827720. Fax: (44) 01235 400454. Lines are open 9.00–5.00, Monday to Saturday, with a 24-hour message answering service. Visit our website at www.hoddereducation.com

First published in 2013 by

Hodder Education, an Hachette UK Company,

338 Euston Road

London NW1 3BH

Impression number   5 4 3 2

Year   2014

Cover photo © senoldo – Fotolia

Illustrations by Datapage (India) Pvt. Ltd

Typeset in 9/11pt Frutiger LT Std 55 Roman by Datapage (India) Pvt. Ltd

Printed and bound by CPI Group (UK) Ltd, Croydon, CR0 4YY

A catalogue record for this title is available from the British Library

ISBN **978 1444 18046 6**

Contents

Contents

Number

1 Number and language

○ Exercises 1.1–1.5

1 List all the prime numbers between 80 and 100.   (2 marks)

....

....

2 List all the factors of the following numbers:

(a) 48 .... (2 marks)

(b) 200 .... (2 marks)

3 List the prime factors of the following numbers and express them as a product of prime numbers:

(a) 25 .... (2 marks)

(b) 48 .... (2 marks)

4 Find the highest common factor of the following numbers:

(a) 51, 68, 85 .... (2 marks)

(b) 36, 72, 108... (2 marks)

5 Find the lowest common multiple of the following numbers:

(a) 8, 12, 16 .... (2 marks)

(b) 2^3, 4^2, 6 .... (2 marks)

○ **Exercise 1.6**

1 State whether each of the following is a rational or irrational number:

(a) 2.5 ....(1 mark)

(b) $0.1\dot{4}$ ....(1 mark)

(c) $\sqrt{17}$ ....(1 mark)

(d) −0.03 ....(1 mark)

(e) $\sqrt{144}$ ....(1 mark)

(f) $\sqrt{5} \times \sqrt{2}$ ....(1 mark)

(g) $\dfrac{\sqrt{16}}{\sqrt{4}}$ ....(1 mark)

2 (a) Draw and name three different solid shapes where the surface area is likely to be a rational number.

(3 marks)

(b) On each of your shapes write on the dimensions which make this true. (Do not work out the surface area). (3 marks)

3 **(a)** Draw two different, compound, two-dimensional shapes (a compound shape is made up of more than one shape) where the total area is likely to be an irrational number.

(2 marks)

   (b) Write on the dimensions of each shape.
       (Do not work out the area.)   **(2 marks)**

4 **(a)** Draw three different solid shapes where the volume is likely to be an irrational number.

(3 marks)

   (b) On each of your shapes write on the dimensions which make this true.
       (Do not work out the volume.)   **(3 marks)**

○ **Exercises 1.7–1.9**

1 Without a calculator work out the following:

(a) $\sqrt{0.81}$ ....(1 mark)

(b) $\sqrt{5\frac{4}{9}}$ ....(2 marks)

2 Without a calculator work out the following:

(a) $\sqrt[3]{-216}$ ....(2 marks)

(b) $\sqrt[3]{15\frac{5}{8}}$ ....(2 marks)

○ **Exercise 1.10**

1 A hang-glider is launched from a mountainside. It climbs 650 m and then starts its descent. It descends 1220 m before landing.

(a) How far below the launch point was the hang-glider when it landed?

....(1 mark)

(b) If the launch point was at 1650 m above sea level, at what height above sea level did it land?

....(1 mark)

2 A plane flying at 8500 m drops a sonar device onto the ocean floor. If the sonar falls a total of 10 200 m, how deep is the ocean at this point?

....(2 marks)

② Accuracy

○ **Exercises 2.1–2.3**

1 Round the following numbers to the nearest 10, 100 or 1000 as shown in brackets:

(a) 47 (10) .... **(1 mark)**

(b) 1250 (100) .... **(1 mark)**

(c) 524 700 (1000) .... **(1 mark)**

2 Write the following numbers to the number of decimal places indicated in brackets:

(a) 4.98 (1 d.p.) .... **(1 mark)**

(b) 18.04 (1 d.p.) .... **(1 mark)**

(c) 0.0048 (2 d.p.) .... **(1 mark)**

3 Write the following numbers to the number of significant figures written in brackets:

(a) 15.01 (1 s.f.) .... **(1 mark)**

(b) 0.042 99 (2 s.f.) .... **(1 mark)**

(c) 3.049 01 (3 s.f.) .... **(1 mark)**

○ **Exercise 2.4**

1 Without using a calculator, estimate the answers to the following calculations:

(a) $\dfrac{47 \times 3.8}{18.8}$ .... **(1 mark)**

(b) $\dfrac{\sqrt{140}}{2.2^2}$ .... **(2 marks)**

2 Estimate the shaded area of the shape below. Do *not* work out an exact answer.

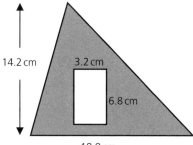

14.2 cm

3.2 cm

6.8 cm

18.8 cm

.... **(3 marks)**

○ **Exercise 2.5**

1 Two pieces of wood measure 14.5 m and 9.4 m, both correct to 1 d.p. What is the lower limit of their total length?

.... (2 marks)

2 A car holds 70 litres of petrol correct to the nearest litre.
Its fuel economy is 12 km per litre to the nearest kilometre.
Write down, but do not work out, the calculation for the upper limit of the distance the car can travel.

....

.... (2 marks)

3 A school has 1500 students correct to the nearest 100.
The cost to run the school is $720 000 to the nearest $10 000.
Write down, but do not work out, the calculation for the lower limit of the cost per student.

....

.... (2 marks)

○ **Exercise 2.6**

1 Calculate upper and lower bounds for the following calculations, if each of the numbers is given to the nearest whole number:

(a) 15×25 ....

.... **(2 marks)**

(b) 128×22 ....

.... **(2 marks)**

(c) 1000×5 ....

.... **(2 marks)**

(d) $\frac{3}{4}$ ....

.... **(2 marks)**

(e) $\frac{120}{60}$ ....

.... **(2 marks)**

2 Calculate upper and lower bounds for the following calculations, if each of the numbers is given to one decimal place:

(a) $2.4 + 14.1$ ....

.... **(2 marks)**

(b) 3.3×8.8 ....

.... **(2 marks)**

(c) 100.0×4.9 ....

.... **(2 marks)**

(d) $21.6 - 12.2$ ....

.... **(2 marks)**

(e) $(0.4 - 0.1)^2$ ....

.... **(2 marks)**

3 If $a = 18$ and $b = 22$, both to the nearest whole number, between what limits is $\sqrt{a^2 + b^2}$?

....

.... **(4 marks)**

○ **Exercise 2.7**

1 A town is to be built on a rectangular plot of land measuring 3.7 km by 5.2 km, both figures correct to 1 d.p. What are the upper and lower limits for the area of the town?

....

....**(2 marks)**

2 303 degrees Kelvin is equivalent to 30 degrees Celsius. Both figures are given to the nearest degree.

(a) What is the maximum percentage error in each case? ....

.... **(2 marks)**

(b) Explain why the percentage errors are different. ....

.... **(1 mark)**

3 1 mile equals 1.6093 km, correct to 4 d.p.
If a distance is 7 miles correct to the nearest mile and 12 km correct to the nearest kilometre, between what are the limits, in kilometres, of the distance?

....

....

....

....**(4 marks)**

Calculations and order

○ **Exercises 3.1–3.2**

1 Represent the inequality $-1 \leqslant x < 4$ on the number line given.

(2 marks)

2 Write the following sentence using inequality signs.
The finishing time (t seconds) of runners in a school 100 m race ranged from 12.1 seconds to 15.8 seconds.

....(1 mark)

3 Write the following decimals in order of magnitude, starting with the smallest:
0.5   0.055   5.005   5.500   0.505   0.550

....(1 mark)

○ **Exercises 3.3–3.5**

1 Using the correct order of operations, calculate the answer to the following without the use of a calculator:

(a) $(25 - 2) \times 10 + 4$

.... (1 mark)

(b) $25 - 2 \times 10 + 4$

.... (1 mark)

(c) $25 - 2 \times (10 + 4)$

.... (1 mark)

2 In the calculations below, insert any brackets that are necessary to make each calculation correct:

(a) $15 \div 3 + 2 \div 2 = 6$ .... (1 mark)

(b) $15 \div 3 + 2 \div 2 = 3.75$ .... (1 mark)

(c) $15 \div 3 + 2 \div 2 = 1.5$ .... (1 mark)

3 Work out the following calculation without a calculator:

$$\frac{8 + 2 \times 4}{4} - 3$$ .... (2 marks)

Integers, fractions, decimals and percentages

○ **Exercises 4.1–4.4**

1 Evaluate the following without a calculator:

    (a) $\frac{3}{8}$ of 32 ....(1 mark)

    (b) $\frac{8}{9}$ of 72 ....(1 mark)

    (c) $\frac{7}{10}$ of 65 .... (2 marks)

2 Change the following mixed numbers to improper fractions:

    (a) $6\frac{3}{5}$ ....(1 mark)

    (b) $3\frac{2}{17}$ ....(1 mark)

3 Without a calculator, change the following improper fractions to mixed numbers:

    (a) $\frac{38}{9}$ ....(1 mark)

    (b) $\frac{231}{15}$ ....(1 mark)

4 Without a calculator, write the following fractions as decimals:

    (a) $3\frac{9}{20}$ ....(1 mark)

    (b) $7\frac{19}{25}$ ....(1 mark)

    (c) $\frac{5}{16}$ .... (2 marks)

5 Without a calculator, complete the table below, giving fractions in their lowest terms.

	Fraction	Decimal	Percentage	
(a)		0.75		(1 mark)
(b)	$\frac{9}{20}$			(1 mark)
(c)			6.5%	(1 mark)
(d)		3.08		(1 mark)
(e)	$\frac{2}{3}$			(1 mark)
(f)		$1.0\dot{5}$		(1 mark)

○ **Exercise 4.5**

1 Work out the following using long division, giving your answer to 2 d.p.:

(a) $4569 \div 12$

(2 marks)

(b) $125 \div 0.13$

(3 marks)

○ **Exercises 4.6–4.10**

1 Evaluate the following without a calculator, leaving your answer as a fraction in its simplest form:

(a) $3\frac{2}{5} - 1\frac{5}{6}$ ....

.... (2 marks)

(b) $\frac{7}{8} - 2\frac{2}{9} + 1\frac{2}{3}$ ....

.... (3 marks)

2 Evaluate the following without a calculator, leaving your answer as a fraction in its simplest form:

(a) $\frac{2}{5} \times 1\frac{2}{9}$ ....

.... (2 marks)

(b) $\left(\frac{4}{9} - 1\frac{4}{5}\right) \div \frac{2}{3}$ ....

.... (3 marks)

3 Change the following fractions to decimals:

(a) $3\frac{4}{9}$ .... (2 marks)

(b) $5\frac{3}{8}$ .... (2 marks)

○ **Exercise 4.11**

1 Convert each of the following recurring decimals to fractions in their simplest form:

(a) $0.5\dot{6}$

....

....

....

....

.... **(2 marks)**

(b) $1.30\dot{8}$

....

....

....

....

.... **(3 marks)**

2 Without using a calculator, evaluate $0.\dot{3}\dot{8} - 0.2\dot{5}$ by converting each decimal to a fraction first.

....

....

....

....

.... **(4 marks)**

Further percentages

○ **Exercises 5.1–5.3**

1 Express the following as percentages:

(a) 0.25 .... (1 mark)

(b) 0.6.. (1 mark)

(c) $\frac{3}{8}$ .... (1 mark)

(d) $\frac{7}{8}$ .... (1 mark)

2 Evaluate the following:

(a) 25% of 200 .... (1 mark)

(b) 75% of 200 .... (1 mark)

(c) $12\frac{1}{2}$% of 400 .... (1 mark)

(d) 130% of $300 .... (1 mark)

(e) 60% of $200 .... (1 mark)

(f) 62.5% of 56.. (2 marks)

3 In a street of 180 houses, 90 of the houses have only one occupant, 45 have two occupants, 36 have three occupants, and the remainder have four or more occupants.

(a) Calculate the percentage of houses with less than four occupants.

....

....

.... (2 marks)

(b) Calculate the percentage of houses with four or more occupants.

....

....

.... (1 mark)

→

4 Simplify each of the following fractions **(i)**, then express them as a percentage **(ii)**:

(a) $\frac{72}{90}$

(i) ....(1 mark)

(ii) ....(1 mark)

(b) $\frac{45}{75}$

(i) ....(1 mark)

(ii) ....(1 mark)

(c) $\frac{26}{39}$

(i) ....(1 mark)

(ii) ....(1 mark)

5 A group of three friends, Ahmet, Jo and Anna, share $180 between them.
Ahmet has $54 of the total, Jo has $81 and Anna the rest. What percentage does each receive?

....

....

....(3 marks)

6 Petrol costs 78.5 cents/litre, and 61 cents of this is tax.
Calculate the percentage that motorists pay in tax.

....(2 marks)

7 Tim buys the following items at a newsagent:

Newspaper	35 cents
Pen	$2.08
Birthday card	$1.45
Sweets	35 cents
Five stamps	29 cents each

(a) If he pays using a $10 note, calculate the amount of change he receives.

....

....(1 mark)

(b) What percentage of the $10 note has he spent?

....

.... (2 marks)

○ **Exercise 5.4**

1 Increase each number by the given percentages:

(a) 180 by 25% ....(1 mark)

(b) 75 by 100% ....(1 mark)

(c) 250 by 250% ....(1 mark)

2 Decrease each number by the given percentages:

(a) 180 by 25% ....(1 mark)

(b) 150 by 30% ....(1 mark)

(c) 8 by 37.5% ....(1 mark)

3 The value of shares in a mobile phone company rises by 135%.

(a) If the value of each share was originally 1620 cents, calculate, to the nearest dollar, the new value of each share.

....

.... (2 marks)

(b) How many shares could now be bought with $10 000?

....

.... (2 marks)

4 During 2012 the average price of a house in London rose by 14%. If the average price of a house was £376 000 at the start of the year, calculate its new value at the end of the year.

....

.... (2 marks)

5 Unemployment figures at the end of last quarter increased by 725 000. If the increase in the number of unemployed this quarter is 7.5% fewer, calculate the increase in the number of people unemployed this quarter.

....

.... (2 marks)

○ **Exercise 5.5**

1 Calculate the value of X in each of the following:

(a) 65% of X is $292.50..(1 mark)

(b) 15% of X is $93.00..(1 mark)

(c) X% of 12 is 40.8..(1 mark)

(d) X% of 20 is 32..(1 mark)

2 In a school 45% of the students are boys. If there are 117 boys in the school, calculate the number of students in the school.

....

....(2 marks)

3 In an exam Paulo scored 68%. If he got 153 marks in total, calculate the number of marks available in the exam.

....

....(2 marks)

4 An elastic band can increase its natural length by 625% when fully stretched. If the elastic band has a length of 29 cm when fully stretched, calculate its natural length.

....

....(3 marks)

Ratio and proportion

◯ Exercise 6.1

1 A bottling machine fills 3000 bottles in one hour.
How many does it fill in a minute?

....(1 mark)

2 A machine prints four sheets of A4 in one minute.
How many does it print in an hour?

....(1 mark)

◯ Exercises 6.2–6.4

1 4g of copper mixes with 5g of tin.

(a) What fraction of the mixture is tin? .... (1 mark)
(b) How much tin is there in 1.8 kg of the same mixture?

.... (1 mark)

2 60% of students in a class are girls.
(a) What is the proportion of girls to boys, in its lowest terms?

.... (1 mark)
(b) What fraction of the same class are boys?

.... (1 mark)
(c) If there are 30 students in the class altogether, how many are girls?

.... (1 mark)

3 A recipe needs 300g of flour to make a dozen cakes.
How many kilograms of flour would be needed to make 100 cakes?

....(1 mark)

4 To make five jam tarts, 80g of jam is needed.
How much jam is needed to make two dozen tarts?

....(1 mark)

→

5 The ratio of the angles of a triangle is 1:2:3.
What is the size of the smallest angle?

....(1 mark)

6 A metre ruler is broken into two parts in the ratio 16:9.
How long is each part?

....(1 mark)

7 A motorbike uses a petrol and oil mixture in the ratio 17:3.
(a) How much of each is there in 25 litres of mixture?

....

.... (2 marks)

(b) How much petrol would be mixed with 250 ml of oil?

.... (1 mark)

8 An aunt gives a brother and two sisters $2500 to be divided in the ratio of their ages.
If the girls are 15 and 17 years old and the boy 18 years old, calculate how much they will
each get.

....

....

....(3 marks)

9 The angles of a hexagon add up to 720° and are in the ratio 1:2:4:4:3:1.
Find the size of the largest and smallest angles.

....

....(2 marks)

10 A company shares profits equally among 120 workers so that they get $500 each.
How much would they each have got had there been 125 workers?

....

....(2 marks)

11 The table below represents the relationship between the speed and the time taken for a train
to travel between two stations.

Speed (km/h)	60			120	90	240
Time (h)	1.5	3	4			

Complete the table. (2 marks)

12 A shop can buy 75 shirts costing $20 each. If the price is reduced by 25%, how many more shirts could be bought?

....

....**(2 marks)**

13 3 people can dig a trench in 30 hours.

   (a) How long would it take:

      (i) 4 people ....**(1 mark)**

      (ii) 5 people? ....**(1 mark)**

   (b) How many people would it take to dig the trench in:

      (i) 15 hours ....**(1 mark)**

      (ii) 45 hours? ....**(1 mark)**

14 A train travelling at 160 km/h takes 5 hours for a journey.
How long would it take a train travelling at 200 km/h?

....**(2 marks)**

15 A swimming pool is filled in 81 hours by 3 identical pumps.
How much quicker would it be filled if 9 similar pumps were used instead?

....

....**(3 marks)**

○ **Exercise 6.5**

1 Increase 250 by the following ratios:

   (a) 8 : 5 ....**(1 mark)**

   (b) 12.5 : 5 ....**(1 mark)**

2 Increase 75 by the following ratios:

   (a) 7.5 : 3 ....**(1 mark)**

   (b) 5 : 2 ....**(1 mark)**

3 Decrease 120 by the following ratios:

   (a) 2 : 3 ....**(1 mark)**

   (b) 1 : 4 ....**(1 mark)**

○ **Exercise 6.6**

1 A photograph measuring 12 cm by 8 cm is enlarged by a ratio of 9 : 4.
What are the dimensions of the new print?

....

....**(2 marks)**

2 A drawing measuring 8 cm by 12 cm needs to be enlarged.
The dimensions of the enlargement need to be 20 cm by 30 cm.
Calculate the enlargement needed and express it as a ratio.

....

....**(2 marks)**

3 A rectangle measuring 24 cm by 12 cm is enlarged by a ratio of 3 : 2.
(a) What is the area of:

    (i) the original rectangle ....**(1 mark)**

    (ii) the enlarged rectangle? ....

    ....**(2 marks)**

(b) By what ratio has the area been enlarged? ....**(1 mark)**

4 A cuboid measuring 12.5 cm by 5 cm by 2.5 cm is enlarged by a ratio of 4 : 1.
(a) What is the volume of:

    (i) the original cuboid ....**(1 mark)**

    (ii) the enlarged cuboid? ....**(2 marks)**

(b) By what ratio has the volume been increased? ....**(1 mark)**

Indices and standard form

○ **Exercises 7.1–7.4**

1 Simplify the following using indices:

(a) $2 \times 2 \times 2 \times 3 \times 3 \times 4 \times 4 \times 4$ .... (1 mark)

(b) $2 \times 2 \times 2 \times 2 \times 4 \times 4 \times 4 \times 4 \times 4 \times 5 \times 5$ .... (1 mark)

(c) $3 \times 3 \times 4 \times 4 \times 4 \times 5 \times 5 \times 5$ .... (1 mark)

(d) $2 \times 7 \times 7 \times 7 \times 7 \times 11 \times 11$ .... (1 mark)

2 Use a calculator to work out the following in full:

(a) 14^2 .... (1 mark)

(b) $3^5 \times 4^3 \times 6^3$ .... (1 mark)

(c) $7^2 \times 8^3$ .... (1 mark)

(d) $13^2 \times 2^3 \times 9^4$ .... (1 mark)

3 Simplify the following using indices:

(a) $11^5 \times 6^3 \times 6^5 \times 6^4 \times 11^2$ .... (1 mark)

(b) $5^4 \times 5^7 \times 6^3 \times 6^2 \times 6^6$ .... (1 mark)

(c) $12^6 \div 12^2$ .... (1 mark)

(d) $13^5 \div 13^2$ .... (1 mark)

4 Simplify the following:

(a) $(9^2)^2$ .... (1 mark)

(b) $(17^2)^5$ .... (1 mark)

(c) $(2^2)^4$ .... (1 mark)

(d) $(8^2)^3$ .... (1 mark)

5 Simplify the following:

(a) $9^2 \times 5^0$ .... (1 mark)

(b) $7^3 \times 7^{-2}$ .... (1 mark)

(c) $16^3 \times 16^{-2} \times 16^{-2}$ .... (2 marks)

(d) $18^0 \div 3^2$ .... (2 marks)

→

6 Work out the following without a calculator:

(a) 2^{-2} .... (2 marks)

(b) 7×10^{-1} .... (2 marks)

(c) 3×10^{-2} .... (2 marks)

(d) 1000×10^{-3} .... (2 marks)

7 Work out the following without a calculator:

(a) 16×2^{-2} .... (2 marks)

(b) 128×2^{-6} .... (2 marks)

(c) 144×6^{-2} .... (2 marks)

(d) $100\,000 \times 10^{-6}$ .... (2 marks)

8 Find the value of x in each of the following:

(a) $2^x = 8$ .... (2 marks)

(b) $4^x = 256$ .... (2 marks)

(c) $10^x = 1\,000\,000$ .... (2 marks)

(d) $5^x = 1$ .... (2 marks)

9 Find the value of z in each of the following:

(a) $2^{(z-1)} = 32$ ....

.... (2 marks)

(b) $3^{(z+2)} = 81$ ....

.... (2 marks)

(c) $4^{2z} = 64$ ....

.... (2 marks)

(d) $2^{-z} = 128^{-1}$ ....

.... (2 marks)

○ **Exercises 7.5–7.6**

1 Write the following numbers in standard form:

(a) 37 000 000 .... (1 mark)

(b) 463 million .... (1 mark)

2 A snail slides at an average speed of 6 cm per minute. Assuming it continues to slide at this rate, calculate how far it travels in centimetres in 24 hours. Write your answer in standard form.

....

.... (2 marks)

3 The Earth has a radius of 6400 km. A satellite 350 km above Earth has a circular path around the Earth as shown in the diagram below.

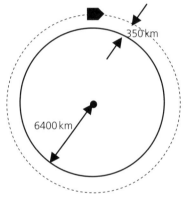

(a) Calculate the radius of the satellite's path. Give your answer in standard form.

.... (1 mark)

(b) Calculate the distance travelled by the satellite in one complete orbit.
Give your answer in standard form correct to one decimal place.

....

.... (2 marks)

4 Write the following numbers in standard form:

(a) 0.000 045 .... (1 mark)

(b) 0.000 000 000 367 .... (1 mark)

5 Deduce the value of x in each of the following:

(a) $0.03^3 = 2.7 \times 10^x$ .... (1 mark)

(b) $0.04^x = 1.024 \times 10^{-7}$ ....

.... (2 marks)

○ **Exercise 7.7**

Evaluate the following without the use of a calculator.

1 $46^{\frac{1}{2}}$ ....(1 mark)

2 $225^{\frac{1}{2}}$ ....(1 mark)

3 $125^{\frac{1}{3}}$ ....(1 mark)

4 $1\,000\,000^{\frac{1}{3}}$ ....(1 mark)

5 $343^{\frac{1}{3}}$ ....(2 marks)

6 $625^{\frac{1}{4}}$ ....(2 marks)

7 $81^{\frac{1}{4}}$ ....(2 marks)

8 $1728^{\frac{1}{3}}$ ....(2 marks)

○ **Exercise 7.8**

Work out the following without the use of a calculator.

1 $\dfrac{17^0}{2^2}$ ....(2 marks)

2 $\dfrac{27^{\frac{2}{3}}}{3^2}$ ....(3 marks)

3 $\dfrac{64^{\frac{1}{2}}}{4^2}$ ....(2 marks)

4 $\dfrac{1^0}{2^3}$ ....(2 marks)

5 $\dfrac{4^{\frac{1}{2}}}{2^2}$ ....(2 marks)

6 $64^{-\frac{1}{2}} \times 2^3$ ....(3 marks)

7 $121^{-\frac{1}{2}} \times 11^2$ ....(3 marks)

8 $729^{-\frac{1}{3}} \div 3^{-2}$ ....(3 marks)

9 $4^{\frac{1}{2}} \times 4^{-2} \times \dfrac{1}{4}$ ....(3 marks)

10 $27^{\frac{1}{3}} \times 81^{-2}$ ....(3 marks)

Money and finance

○ Exercise 8.1

1   The table below shows the exchange rate for €1 into various currencies.

Brazil	2.6 Brazilian reals
China	8.0 Chinese yuans
New Zealand	1.5 New Zealand dollars
Sri Lanka	162 Sri Lanka rupees

Convert the following:

(a) 150 Brazilian reals to euros ....(1 mark)

(b) 1000 Sri Lanka rupees to euros ....(1 mark)

(c) 500 Chinese yuan to New Zealand dollars ....

.... (3 marks)

○ Exercises 8.2–8.4

1   Manuela makes different items of pottery. The table below shows the number of each item she makes and the amount she is paid for each item.

Item	Amount paid per item	Number made
Cup	€2.30	15
Saucer	€0.75	15
Teapot	€12.25	3
Milk jug	€3.50	6

(a) Calculate her gross earnings. ....

.... (1 mark)

(b) Tax deductions are 18% of gross earnings. Calculate her net pay.

....

.... (3 marks)

→

2 A caravan is priced at $9500. The caravan supplier offers customers two different options for buying the caravan. They are as follows:

Option 1: A deposit of 25% followed by 24 monthly payments of $350

Option 2: 36 monthly payments of $380

(a) Calculate the amount extra a customer would have to pay with each of the options.

....

....

.... **(3 marks)**

(b) Explain why a customer might want to choose the more expensive option.

....

.... **(2 marks)**

3 A professional baker makes cakes. The ingredients for each cake cost the baker $3.80. If he sells each cake for $9.20, calculate his percentage profit.

....

.... **(2 marks)**

4 A house is bought for $240 000. After 5 years its value has decreased to $180 000. Calculate the average yearly percentage depreciation.

....

....

.... **(3 marks)**

○ **Exercise 8.5**

1 What simple rate of interest is paid on a deposit of $5000 if it earns $200 interest in 4 years?

....

.... **(2 marks)**

2 How long will it take a principal of $800 to earn $112 of simple interest at 2% per year?

....

.... **(2 marks)**

○ **Exercise 8.6**

1 A couple borrow $140 000 to buy a house at 5% compound interest for three years. How much will they pay at the end of the three years?

....

....**(3 marks)**

2 A man buys a BMW for $50 000.
He pays with a loan at 10% compound interest for three years. What did his BMW cost him?

....

....**(3 marks)**

3 A girl owes $250 on a credit card. The APR is 20%. What does she owe in four years if she pays nothing back?

....

....**(3 marks)**

4 In five years a debt has doubled. What was the compound interest?

....

....

....**(4 marks)**

5 A boat has halved in value in three years. What was the percentage loss in compound terms?

....

....

....**(4 marks)**

6 A internet company grows by 20% each year.
(a) Explain why it will not take 5 years to double in size.

....

....**(2 marks)**

(b) When will it double in size?

....

....**(4 marks)**

Time

○ Exercise 9.1

1 A cyclist sets off on a ride at 09 25. If his journey takes 327 minutes, calculate the time he finishes cycling.

   ....

   .... **(2 marks)**

2 A plane travels 7050 km at an average speed of 940 km/h. If it lands at 13 21, calculate the time it departed.

   ....

   .... **(3 marks)**

3 A train travelling from Paris to Istanbul departs at 16 30 on a Wednesday. During the journey it stops at several locations. Overall the train travels the 2280 km distance at an average speed of 18 km/h.

   (a) Calculate the time taken to travel to Istanbul.

   ....

   .... **(2 marks)**

   (b) What day of the week does the train arrive in Istanbul?

   .... **(1 mark)**

   (c) What time of the day does the train arrive in Istanbul? Give your answer to the nearest minute.

   ....

   .... **(3 marks)**

Set notation and Venn diagrams

○ **Exercise 10.1**

1 **(a)** Describe the following set in words:
{Moscow, London, Cairo, New Delhi, …}

....(1 mark)

(b) Write down two more elements of this set.

....(2 marks)

2 **(a)** Describe the following set in words:
{euro, dollar, yen,… }

....(1 mark)

(b) Write down two more elements of this set.

....(2 marks)

3 Consider the set $P = \{(x, y): y = x^2 + x\}$.
Write down two elements of the set.

....(2 marks)

4 Consider the set $R = \{p: -1 \leqslant p < 7\}$.
(a) Describe the set.

....(1 mark)

(b) Write down two elements of the set.

....(2 marks)

○ **Exercise 10.2**

1 The set $A = \{x: 0 < x < 10\}$.
(a) List the subset B {prime numbers}.

....(2 marks)

(b) List the subset C {square numbers}.

....(2 marks)

→

2 $P = \{a, b, c\}$

(a) List all the subsets of P.

....

.... (2 marks)

(b) List all the proper subsets of P.

....

.... (1 mark)

○ Exercise 10.3

1 If $\mathscr{E} = \{$girl's names$\}$ and $M = \{$girl's names beginning with the letter A$\}$ what is the set represented by M'?

.... (1 mark)

2 The Venn diagram below shows the relationship between three sets of numbers A, B and C.

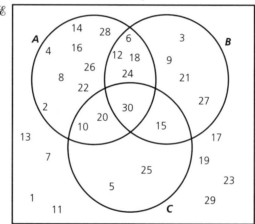

(a) If $\mathscr{E} = \{1, 2, 3, 4 \dots 30\}$, complete the following statements:

(i) $A = $ .... (1 mark)

(ii) $B = $ .... (1 mark)

(iii) $C = $ .... (1 mark)

(b) Complete the following by entering the correct numbers:

(i) $A \cap B = \{$ .... $\}$ (1 mark)

(ii) $B \cup C = \{$ .... $\}$ (1 mark)

(iii) $A \cap B \cap C = \{$ .... $\}$ (1 mark)

(iv) $A' \cap C = \{$ .... $\}$ (2 marks)

3   Consider the Venn diagram below showing the relationship between the sets *W*, *X*, *Y* and *Z*.

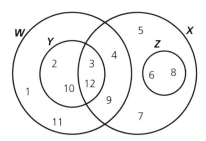

(a) Complete the following by entering the correct numbers:

(i) $X \cup Y =$ .... (1 mark)

(ii) $W \cap X =$ .... (1 mark)

(b) Which of the named sets is a subset of *X*? .... (1 mark)

○ **Exercise 10.4**

1   The sets given below represent the letters of the alphabet in each of three English cities.
$P = \{c,a,m,b,r,i,d,g,e\}$, $Q = \{b,r,i,g,h,t,o,n\}$ and $R = \{d,u,r,h,a,m\}$

(a) Draw a Venn diagram to illustrate this information.

(3 marks)

(b) Complete the following statements:

(i) $Q \cup R = \{$...$\}$ (1 mark)

(ii) $P \cap Q \cap R = \{$...$\}$ (1 mark)

(iii) $Q = P' = \{$...$\}$ (2 marks)

→

2   In a class of 30 students, 16 do athletics (*A*), 17 do swimming (*S*), whilst 3 do neither.

(a) Complete the Venn diagram below.

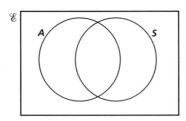

(3 marks)

(b) Calculate the following:

(i) $n(A \cap S)$ ....(1 mark)

(ii) $n(A \cup S)'$ ....(1 mark)

○ **Exercise 10.5**

1   A class of 15 students was asked what pets they had. Each student had either a dog (*D*), cat (*C*), fish (*F*), or a combination of them.

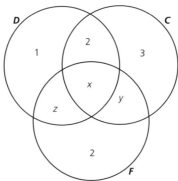

(a) If $n(D) = 10$, $n(C) = 11$ and $n(F) = 9$, calculate:

(i) *z*..

.... (3 marks)

(ii) *x*...

.... (2 marks)

(iii) *y*...

....(1 mark)

(b) Calculate $n(C \cup F)$. ....

....(1 mark)

○ **Exam focus**

1 State whether the following number is rational or irrational. Justify your answer.

$1+\left(3\times\sqrt{81}\right)$

....[2]

2 A square has an area of 225 cm² to the nearest whole number.
   (a) What is the maximum area of the square?

....[1]

   (b) What are the maximum dimensions of the square to 2 d.p.?

....[1]

   (c) What is the minimum perimeter of the square to 2 d.p.?

....[2]

3 **(a)** Express 64% as a fraction in its lowest terms. ....[1]
   (b) Find 64% of 350. ....

....[1]

   (c) Luca score 64% in a maths exam. He got 160 marks.
   How many marks were available in the exam?

....[2]

   (d) Luca needed 78% or more for an A grade. How many more marks did he need?

....[2]

4 The side length of cube *A* and side length of cube *B* are in the ratio 3:2.
   What is the ratio of their volumes?

....[3]

5 Evaluate the following without using a calculator:

   (a) $\dfrac{625^{-\frac{1}{2}}}{5}\times 5^2\times\dfrac{1}{5}$ ....

....[3]

   (b) $\dfrac{64^{\frac{1}{3}}+5}{3^2}$ ....

....[3]

→

6 (a) Use a calculator to work out the following:

(i) $\left(\dfrac{1}{2}\right)^{\frac{1}{2}}$ ....[1]

(ii) $\left(\dfrac{3}{2}\right)^{\frac{3}{2}}$ ....[1]

(iii) $\left(\dfrac{5}{2}\right)^{\frac{5}{2}}$ ....[1]

(b) Using trial and error, solve the following. Give your answer correct to 1 d.p.
$x^x = 1000$

....

....

....

....[2]

7 A water molecule is made up of two hydrogen atoms and one oxygen atom. The mass of a water molecule is 3.0×10^{-26} kg. If the mass of a hydrogen atom is 1.674×10^{-27} kg, calculate the mass of an oxygen atom.

....

....[3]

8 Octay is thinking of buying a caravan for $20 000. He knows that it will lose 30% of its value in the first year and 25% in the second year.

(a) What is it worth at the end of year 1? ....[1]

(b) What is it worth at the end of the second year? ....[1]

(c) How much has he lost in depreciation? ....[1]

(d) Octay could have invested his money at 4% compound interest. What money would he have had after 2 years?

....

....[3]

9 The sets *P*, *Q* and *R* are shown in the Venn diagram below.

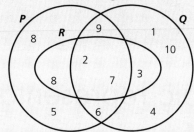

(a) Complete the following statements:

(i) $R = \{$...$\}$ [1]

(ii) $R \cap Q = \{$..$\}$ [1]

(iii) $P \cap Q \cap R' = \{$...$\}$ [2]

(b) Is *R* a subset of *P*? Justify your answer.

....[2]

Algebra and graphs

(11) Algebraic representation and manipulation

○ **Exercises 11.1–11.3**

Expand the following and simplify where possible.

1 $-5(x + 4)$ .... (1 mark)

2 $-3(y - 2)$ .... (1 mark)

3 $4a(2b + 4)$ .... (1 mark)

4 $6(2c - 8)$ .... (1 mark)

5 $-3a^2(2a - 3b)$ ....

....(2 marks)

6 $12(p + 3) - 12(p - 1)$ ....

....(2 marks)

7 $5a(a + 3) - 5(a^2 - 1)$ ....

....(2 marks)

8 $\frac{1}{2}(8x + 4) + 2(3x + 6)$ ....

....(2 marks)

9 $2(2x + 6y) + \frac{3}{4}(4x - 8y)$ ....

....(2 marks)

10 $\frac{1}{8}(16x - 24y) + 4(x - 5y)$ ....

....(2 marks)

Expand and simplify the following:

11 $4p - 3(p + 7)$ ....

....(1 mark)

12 $3q(2 + 7r) + 2r(3 + 4q)$ ....

....

....(2 marks)

13 $-2x(2y - 3z) - 2y(2z - 2y)$ ....

....

....(2 marks)

14 $\frac{a}{9}(27 + 72b)$ ....

....

....(2 marks)

15 $\frac{p}{2}(4q - 4) - \frac{p}{3}(9q - 9)$ ....

....

....(2 marks)

16 $(a + 8)(a + 4)$ ....

....(2 marks)

17 $(b - 3)(b + 3)$ ....

....(2 marks)

18 $(c - 9)(c - 9)$ ....

....(2 marks)

19 $(1 - m)(1 - m)$ ....

....(2 marks)

20 $(j + k)(k - m)$ ....

....(2 marks)

◯ Exercise 11.4

Factorise the following.

1 $3a + 6b$ ....(1 mark)

2 $-14c - 28d$ ....(1 mark)

3 $42x^2 - 21xy^2$ ....

....(2 marks)

4 $m^3 - m^2n - n^2m$ ....

....(2 marks)

5 $-13p^2 - 32r^3$ ....

....(2 marks)

○ Exercise 11.5

Evaluate the expressions below if: $p = 3$, $q = -3$, $r = -1$ and $s = 5$.

1 $p - q + r - s$ ....

.... (2 marks)

2 $5(p + q + r + s)$ .... (2 marks)

.... (2 marks)

3 $2p(q - r)$ .... (2 marks)

.... (2 marks)

4 $p^2 + q^2 + r^2 + s$ ....

.... (2 marks)

5 $-p^3 - q^3 - r^3 - s^3$ ....

.... (2 marks)

○ Exercise 11.6

Make the letter in bold the subject of the formula.

1 $ab + \mathbf{c} = d$ ....

....

.... (2 marks)

2 $a\mathbf{b} - c = d$ ....

....

.... (2 marks)

3 $\frac{1}{8}\mathbf{m} + 3 = 2r$ ....

....

.... (2 marks)

4 $p - \frac{\mathbf{q}}{r} = s$ ....

....

.... (2 marks)

5 $\frac{p}{-\mathbf{q}} + r = -s$ ....

....

.... (2 marks)

◯ **Exercise 11.7**

Expand and simplify the following.

1 $(2d + 3)(2d - 3)$ ....

....**(2 marks)**

2 $(3e - 7)(3e - 7)$ ....

....**(2 marks)**

3 $(2f + 3g)(2f - 3g)$ ....

....**(2 marks)**

4 $(4 - 5h)(5h + 4)$ ....

....

....**(2 marks)**

5 $(2x + 1)(3x - 1)$ ....

....

....**(2 marks)**

◯ **Exercise 11.8**

Factorise the following by grouping.

1 $ac + a + b + bc$ ....

....

....**(2 marks)**

2 $3cd + 3d + 4e + 4ce$ ....

....

....**(2 marks)**

3 $fg - 4f - 6g + 24$ ....

....

....**(2 marks)**

4 $p^2 - 2pq - 2pr + 4rq$ ....

....

....**(2 marks)**

5 $16m^2 + 44mn + 121n + 44m$ ....

....

....**(2 marks)**

○ **Exercise 11.9**

Factorise the following.

1 $16m^2 - 121n^2$ ....

....(2 marks)

2 $x^6 - y^6$ ....

....(2 marks)

3 $9a^4 - 144b^4$ ....

....(2 marks)

4 $81m^2 - 16n^2$ ....

....(2 marks)

○ **Exercise 11.10**

By factorising, evaluate the following.

1 $17^2 - 16^2$ ....

....(2 marks)

2 $3^4 - 1$ ....

....(2 marks)

3 $98^2 - 4$ ....

....(2 marks)

○ **Exercise 11.11**

Factorise the following quadratic expressions.

1 $a^2 + 5a + 6$ ....

....(2 marks)

2 $b^2 - 3b - 10$ ....

....(2 marks)

3 $c^2 - 10c + 16$ ....

....(2 marks)

4 $d^2 - 18d + 81$ ....

.... **(2 marks)**

5 $2e^2 + 3e + 1$ ....

.... **(3 marks)**

6 $3f^2 + f - 2$ ....

.... **(3 marks)**

7 $2g^2 - g - 1$ ....

.... **(3 marks)**

8 $9h^2 - 4$ ....

.... **(3 marks)**

9 $j^2 + 4jk + 4k^2$ ....

.... **(3 marks)**

○ Exercises 11.12–11.13

In the formulas below make 'a' the subject.

1 $\dfrac{p}{q} = \dfrac{2xa}{r}$

....

.... **(3 marks)**

2 $\dfrac{ma^2}{3n} = \dfrac{2}{n}$

....

.... **(3 marks)**

3 $t = \dfrac{2r\sqrt{a}}{b}$

....

.... **(3 marks)**

4 $t = \dfrac{2p\sqrt{b}}{a}$

....

.... **(3 marks)**

5 $\dfrac{2\sqrt{a}}{3} = \dfrac{b^2}{c}$

....

.... **(3 marks)**

○ **Exercise 11.14**

In each of the questions below:
   (a) change the subject of the formula
   (b) solve the problem.

1 The circumference of a circle is given by the equation $C = 2\pi r$.
Find r when C is 18.8 cm.

....(2 marks)

....(2 marks)

2 The area of a circle is given by the equation $A = \pi r^2$.
Find r when $A = 78.5$ cm^2.

....(2 marks)

....(2 marks)

3 A parallelogram has area 48 cm^2 and one side 8 cm. Find the perpendicular height of the parallelogram using the formula $A = lp$, where 'l' is the length of a side and 'p' is the perpendicular height.

....(2 marks)

....(2 marks)

4 The surface area of a cylinder is 188 cm^2. Its radius is 3 cm. Given the formula $A = 2\pi r(r + h)$, rearrange to find an expression for h, then find the value of h. Draw a sketch if necessary.

....(2 marks)

....(2 marks)

5 A cylinder of radius 4 cm has a volume of 503 cm^3. Given the formula $V = \pi r^2 h$, rearrange to find an expression for h, then find the value of h. Draw a sketch if necessary.

....(2 marks)

....(2 marks)

○ **Exercise 11.15**

Simplify the following fractions.

1 $\dfrac{2a^2}{3} \times \dfrac{6}{a}$ ....

....(2 marks)

2 $\dfrac{5c^2}{2d} \times \dfrac{6e}{c} \times \dfrac{d^2}{c}$ ....

....(2 marks)

3 $\dfrac{9p}{7} \times \dfrac{14}{3p}$ ....

....(2 marks)

4 $\dfrac{6}{r} \times \dfrac{5d}{3s} \times \dfrac{3rs}{2d}$ ....

....(2 marks)

5 $\dfrac{4x^3}{3y^4} \times \dfrac{6y^5}{2x^2}$ ....

....(2 marks)

○ **Exercises 11.16–11.17**

Simplify the following fractions.

1 $\dfrac{a}{4} + \dfrac{b}{3}$ ....

....

....(2 marks)

2 $\dfrac{3c}{4} - \dfrac{2c}{3}$ ....

....

....(2 marks)

3 $\dfrac{a}{2} + \dfrac{a}{3}$ ....

....

....(2 marks)

4 $\dfrac{3e}{7} - \dfrac{2e}{3}$ ....

....

....(2 marks)

5 $\dfrac{f}{9} + f$ ....

....

....(2 marks)

○ **Exercise 11.18**

Simplify the following fractions.

1 $\dfrac{1}{p+3} + \dfrac{2}{p-1}$ ....

....

....

....**(3 marks)**

2 $\dfrac{a(a+5)}{b(a+5)}$ ....

....

....

....**(3 marks)**

3 $\dfrac{a^2 - 3a}{(a+1)(a-3)}$ ....

....

....

....**(3 marks)**

4 $\dfrac{a^2 + 2a}{a^2 + 5a + 6}$ ....

....

....

....**(3 marks)**

5 $\dfrac{a^3 - a}{a^2 - 1}$ ....

....

....

....**(3 marks)**

(12) Algebraic indices

○ Exercises 12.1–12.2

1 Simplify the following using indices:

   (a) $a^5 \times a^3 \times b^5 \times b^4 \times c^2$ .... (2 marks)

   (b) $p^4 \times q^7 \times p^3 \times q^2 \times r$ .... (2 marks)

   (c) $m^9 \div m^2 \div (m^2)^4 \times m^2$ .... (2 marks)

   (d) $a^5 \times e^3 \times b^5 \times e^4 \times e^2 \times e^5 \div e^{13}$ .... (2 marks)

2 Simplify the following:

   (a) $ac^5 \times ac^3$ .... (2 marks)

   (b) $m^4n \div nm^2$ .... (2 marks)

   (c) $(b^3)^3 \div b^8$ .... (2 marks)

   (d) $3(2b^3)^3$ .... (2 marks)

○ Exercise 12.3

1 Rewrite the following in the form $a^{\frac{m}{n}}$:

   (a) $\left(\sqrt[4]{a}\right)^5$ .... (2 marks)

   (b) $\left(\sqrt{a}\right)^7$ .... (2 marks)

2 Rewrite the following in the form $\left(\sqrt[n]{b}\right)^m$:

   (a) $b^{-\frac{3}{5}}$ .... (2 marks)

   (b) $b^{\frac{7}{9}}$ .... (2 marks)

3 Simplify the following algebraic expressions, giving your answer in the form $a^{\frac{m}{n}}$:

   (a) $a^{\frac{2}{3}} \times a^{-\frac{3}{4}}$ .... (2 marks)

   (b) $\dfrac{a^{-3}}{\sqrt[3]{a}}$ .... (2 marks)

   (c) $\dfrac{(a^{-2})^3}{a^{-\frac{5}{2}} \times \left(\sqrt[3]{a}\right)^{-2}}$ .... (3 marks)

(13) Equations and inequalities

○ Exercise 13.1

Solve the following linear equations.

1 $4a = 12 + 3a$ ....(1 mark)

2 $5 = 17 + 4b$ ....(1 mark)

3 $3c - 9 = 5c + 13$ ....

....(2 marks)

4 $\dfrac{d}{7} = 2$ ....

....(1 mark)

5 $\dfrac{e}{3} - 2 = 4$ ....

....(2 marks)

6 $\dfrac{3f}{5} - 1 = 5$ ....

....(2 marks)

7 $\dfrac{2g - 1}{3} = 3$ ....

....(2 marks)

8 $\dfrac{4(h + 5)}{3} = 12$ ....

....(2 marks)

9 $\dfrac{7 - 2j}{5} = \dfrac{11 - 3j}{8}$ ....

....

....(3 marks)

10 $3(2k + 4) = 2(5k - 4)$ ....

....

....(3 marks)

○ **Exercise 13.2**

1 The triangle below has angles $x°$, $x°$ and $(x + 30)°$. Find the value of each angle.

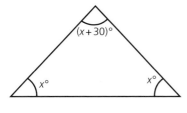

....(3 marks)

2 The triangle below has angles $x°$, $(x + 40)°$ and $(2x - 20)°$ degrees.
Find the value of each angle.

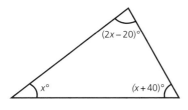

....(3 marks)

3 The isosceles triangle below has its equal sides of length $(3x + 20)$ cm and $(4x - 5)$ cm.
Calculate the value of x.

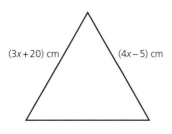

....(3 marks)

4 Two straight lines cross at opposite angles of $(7x + 4)°$ and $(9x - 32)°$ degrees as shown.
Calculate the size of all four angles.

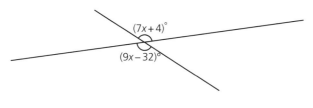

....(3 marks)

5 The area of a rectangle is 432 cm². Its length is three times its width. Draw a diagram and work out the size of the sides.

....(3 marks)

Calculate the angles in the following.

6

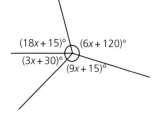

....(4 marks)

7

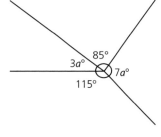

....(4 marks)

8

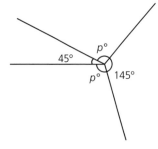

....(3 marks)

9

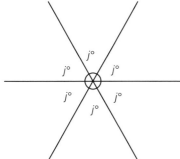

....(2 marks)

10

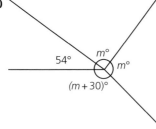

....(3 marks)

11 A right–angled triangle has two acute angles of $(4x - 45)°$ and $(9x - 60)°$.
Calculate their size in degrees.

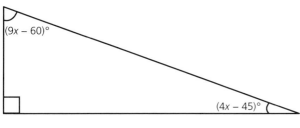

....(3 marks)

12 The interior angles of a regular pentagon add up to 540 degrees.
A pentagon has angles $(4x + 20)°$, $(x + 40)°$, $(3x − 50)°$, $(3x − 130)°$ and 110° as shown. Find the value of each angle.

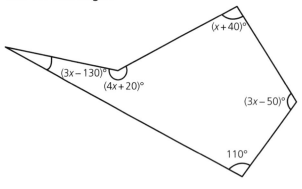

....(3 marks)

13 An isosceles trapezium has angles as shown. Find the value of x.

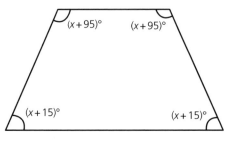

....(3 marks)

○ **Exercises 13.3–13.5**

Solve the following simultaneous equations.

1 $a + b = 12$
$a − b = 2$

....

....(2 marks)

2 $3c + d = 19$
$3c + 4d = 49$

....

....(3 marks)

3 $7e + 4f = 56$

$e + 4f = 32$

....

....**(3 marks)**

4 $g + h = -12$

$g - h = 2$

....

....**(2 marks)**

5 $-5p - 3q = -24$

$-5p + 3q = -6$

....

....**(3 marks)**

6 $2r - 3s = 0$

$2r + 4s = -14$

....

....**(3 marks)**

7 $w + x = 0$

$w - x = 10$

....

....**(2 marks)**

8 $x + y = 2$

$x - y = 1$

....

....**(2 marks)**

9 $2a + 3b = 12$

$a + b = 5$

....

....**(3 marks)**

10 $3c - 3d = 12$

$2c + d = 11$

....

....**(3 marks)**

→

11 $e - f = 0$

$4e + 2f = -6$

....

....**(3 marks)**

12 $12g + 6y = 15$

$g + 2y = 2$

....

....**(3 marks)**

13 $4h + j = 14$

$12h - 6j = 6$

....

....**(4 marks)**

14 $100k - 10l = -20$

$-15k + 3l = 9$

....

....**(4 marks)**

15 $-3 = m + n$

$m - n = 11$

....

....**(3 marks)**

16 $3 - p = q$

$3 - q = 2$

....

....**(3 marks)**

17 $3r - 2s = 26$

$4s + 2 = r$

....

....**(4 marks)**

18 $\frac{1}{2}t + 2w = 1$

$4w - t = 0$

....

....**(4 marks)**

19 The sum of two numbers is 37 and their difference is 11. Find the numbers.

....

....**(3 marks)**

20 The sum of two numbers is −2 and their difference is 12. Find the numbers.

....

....**(3 marks)**

21 If a girl multiplies her age in years by four and adds three times her brother's age, she gets 64. If the boy adds his age in years to double his sister's age, he gets 28. How old are they?

....

....

....**(4 marks)**

22 A rectangle has opposite sides of $3a + b$ and 25 and $2a + 3b$ and 26 as shown. Find the values of a and b.

....**(4 marks)**

23 A square has sides $2x$, $40 - 3x$, $25 + 3y$ and $10 - 2y$. Calculate:

(a) the values of x and y ....

....**(2 marks)**

(b) the area of the square ....

....**(3 marks)**

(c) the perimeter of the square ....

....**(1 mark)**

24 A grandmother is four times as old as her granddaughter. She is also 48 years older than her. How old are they both?

....

....

....**(3 marks)**

○ **Exercise 13.6**

1   A number is trebled then four is added. The total is –17. Find the number.

....

....

....(2 marks)

2   Two is the answer when 20 is added to three times a number. Find the number.

....

....

....(2 marks)

3   A number divided by 17 gives –4. Find the number.

....

....

....(2 marks)

4   A number squared, divided by 5, less 1, is 44. Find two possible values for the number.

....

....

....(5 marks)

5   Zach is two years older than his sister Leda and three years younger than his dog, Spot.
    (a) By writing Zach's age as *x*, write expressions for the ages of Leda and Spot in terms of *x*.

    ....

    .... (2 marks)

    (b) Find their ages if their total age is 22 years.

    ....

    .... (2 marks)

6   A decagon has five equal exterior angles, whilst the others are three times bigger.
    Find the size of the two different angles.

....

....

....

....(4 marks)

7 A triangle has interior angles of $x°$, $2x°$, $6x°$ as shown.

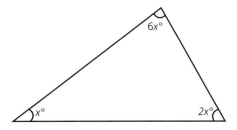

Find the size of its **exterior** angles.

....

....**(4 marks)**

8 A number squared has the number squared then doubled added to it. The total is 300.
Find two possible values for the number.

....

....

....

....**(4 marks)**

○ **Exercise 13.7**

Solve the following equations and give two solutions for x.

1 $x^2 + x - 12 = 0$

....

....

....**(3 marks)**

2 $x^2 - 9x + 18 = 0$

....

....

....**(3 marks)**

3 $x^2 + 10x + 21 = 0$

....

....

....**(3 marks)**

→

4 $x^2 = -(3x + 2)$

....

....

....**(3 marks)**

5 $x^2 - 2x = 35$

....

....

....**(3 marks)**

6 $-42 + 13x = x^2$

....

....

....**(3 marks)**

7 $x^2 - 169 = 0$

....

....

....**(3 marks)**

8 $x^2 - 40 = 9$

....

....

....**(3 marks)**

○ Exercise 13.8

Solve the following quadratic equations where possible.

1 $2x^2 + 8x + 6 = 0$

....

....

....**(4 marks)**

2 $3x^2 + 4x = -1$

....

....

....**(4 marks)**

3 $5x^2 = 4x + 1$

....
....
....**(4 marks)**

4 $3x^2 = 108$

....
....
....**(3 marks)**

5 $3x^2 = 27$

....
....
....**(3 marks)**

6 $3x^2 = -36$

....
....
....**(3 marks)**

7 $3x^2 = -108$

....
....
....**(3 marks)**

8 $4x^2 = 1$

....
....
....**(3 marks)**

9 $16x^2 = 1$

....
....
....**(3 marks)**

10 $3x^2 = \dfrac{4}{3}$

....
....
....**(3 marks)**

→

11 $25x^2 = 64$

....

....

....**(3 marks)**

12 $16x^2 = -64$

....

....

....**(3 marks)**

○ **Exercise 13.9**

1 I have a number of dollars in my pocket. If I square the amount, it is the same as 21 dollars more than four times the amount. How much do I have?

....

....

....**(4 marks)**

2 A triangle has base length $2b$ cm and a height two less than $2b$ cm. Its area is $60\,cm^2$ as shown. What is the base length and height?

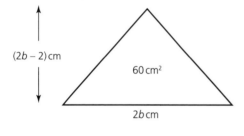

$(2b - 2)$ cm

$60\,cm^2$

$2b$ cm

....**(4 marks)**

3 Drawn below is a right–angled triangle with a hypotenuse of 13 cm. The two shorter sides are x cm and $x + 7$ cm long. If the square of the longest side equals the sum of the squares of the other two sides (Pythagoras' theorem), find the length of the two sides.

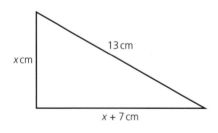

13 cm

x cm

$x + 7$ cm

....**(4 marks)**

4 Two consecutive numbers multiply to make 552. Find the numbers.

....

....

....(4 marks)

5 A man buys a number of golf balls for \$6. If he had paid 50 cents less for each, he could have bought six more for \$6. How many balls did he buy?

....

....

....(4 marks)

○ **Exercise 13.10**

Solve the following quadratic equations either by using the quadratic formula or by completing the square.

1 $6x^2 + 22x = -12$

....

....

....(4 marks)

2 $x^2 = -(1 + 4x)$

....

....

....(4 marks)

3 $10x - 2 = -4x^2$

....

....

....

....(4 marks)

4 $3 - x = 3x^2$

....

....

....

....(4 marks)

→

5 $15 = 7x + 2x^2$

....

....

....

....**(4 marks)**

6 $4y^2 = -5y - 1$

....

....

....

....**(4 marks)**

7 $8x - 4x^2 = -6$

....

....

....

....**(4 marks)**

8 $10x^2 = 60 - 25x$

....

....

....

....**(4 marks)**

9 $2x^2 + 6.6x - 1.4 = 0$

....

....

....

....**(4 marks)**

10 $2x(x + 1) = x^2 - 2x - 4$

....

....

....

....**(4 marks)**

○ **Exercises 13.11–13.12**

Write the following as linear inequalities using the correct mathematical symbols and show the solution on a number line.

1 16 plus $2x$ is less than 10...

(3 marks)

2 19 is greater than or equal to $9x$ plus 1..

(3 marks)

3 1 minus $3x$ is equal to or exceeds 13 ....

(3 marks)

4 A half x is smaller than 2 ....

(3 marks)

5 A third of x is equal to or bigger than 1 ....

(3 marks)

6 $4x$ is more than 8 but less than 16 ....

(4 marks)

7 $9x$ is between 9 and 45 but not equal to either ....

(4 marks)

8 $2x - 6$ is between 4 and 10 but equal to neither...

(4 marks)

9 3 is equal to or is less than $2x + 1$ which is less than 9 ....

(4 marks)

10 20 is equal to or bigger than $2x - 5$ which is greater than or equal to 10.

.... (4 marks)

Linear programming

○ Exercise 14.1

Solve each of the following inequalities.

1 $4x - 12 \geqslant -6$ ....

....**(1 mark)**

2 $2 \leqslant -3x + 8$ ....

....**(1 mark)**

3 $9 < -3(y + 4) \leqslant 15$ ....

....**(2 marks)**

○ Exercise 14.2

In each question below, shade the region which satisfies the inequality, on the axes given.

1 $y \leqslant \frac{1}{2}$

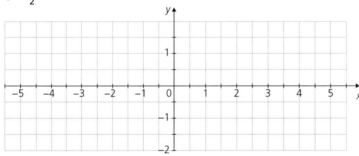

(1 mark)

2 $y + 2x - 4 < 0$

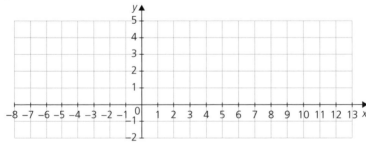

(2 marks)

3 $x - 3y \geqslant 6$

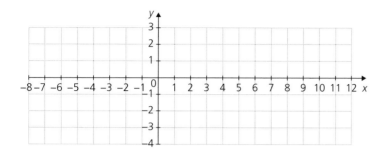

(3 marks)

⃝ **Exercise 14.3**

On the same pair of axes plot the following inequalities and leave **unshaded** the region which satisfies all of them simultaneously.

1 $y \leqslant -2x - 3$, $y > -\frac{1}{2}x - 3$, $x \geqslant -3$

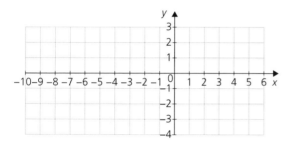

(3 marks)

2 $y \leqslant -\frac{1}{2}x + 2$, $y \geqslant 0$, $3y + 2x - 6 > 0$

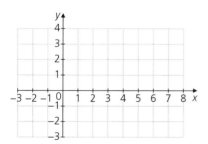

(3 marks)

○ **Exercise 14.4**

1 A team at the Olympics has the following number of male (x) and female (y) athletes:

- the number of male athletes is greater than 5
- the number of female athletes is greater than 7
- the total number of athletes is less than or equal to 15.

(a) Express each of the three statements above as inequalities.

....

....

.... **(3 marks)**

(b) On the axes below, identify the region that satisfies all the inequalities by shading the unwanted regions.

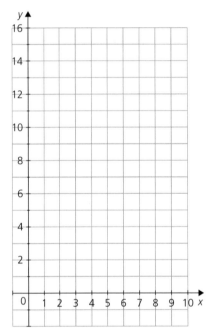

(6 marks)

(c) State the possible solution(s) for the number of male and female athletes in this Olympic team.

....

....

.... **(3 marks)**

(15) Sequences

○ Exercises 15.1–15.2

Give the next two terms in each of the following sequences of numbers.

1 17, 20, 23, 26 ....(1 mark)

2 2, 5, 10, 17 ....(1 mark)

3 5, 13, 21, 29 ....(1 mark)

4 **(a)** Draw the next two patterns in the sequence below.

 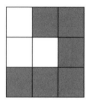 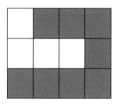

(2 marks)

(b) Complete the table below linking the number of white squares to the number of shaded squares.

Number of white squares	2	3	4	5	6
Number of shaded squares					

(2 marks)

(c) Write the rule for the nth term of the sequence.

.... (2 marks)

(d) Use your rule to predict the number of shaded squares in a pattern with 50 white squares.

.... (2 marks)

→

5 For each of the sequences given below:

    (i) calculate the next two terms

    (ii) explain the pattern in words.

    (a) 9, 16, 25, 36 .... **(2 marks)**

    .... **(2 marks)**

    (b) 12, 24, 36, 48 .... **(2 marks)**

    .... **(2 marks)**

    (c) 1, 1, 2, 3, 5, 8, 13 .... **(2 marks)**

    .... **(2 marks)**

6 For each of the sequences shown below, give an expression for the nth term:

    (a) 7, 11, 15, 19 .... **(2 marks)**

    (b) 7, 9, 11, 13 .... **(2 marks)**

    (c) 3, 6, 11, 18 .... **(2 marks)**

○ **Exercise 15.3**

Using a table if necessary:

    (a) give the next two terms in each of the following sequence

    (b) find the formula for the nth term.

1 0, 7, 26, 63, 124

    (a) ....

    ....

    .... **(2 marks)**

    (b) ....

    .... **(2 marks)**

2 3, 10, 29, 66, 127

    (a) ....

    ....

    .... **(3 marks)**

    (b) ....

    .... **(3 marks)**

○ **Exercise 15.4**

Give the next two terms in each of the following sequences of numbers.

1 64, 32, 16, 8 .... **(1 mark)**

2 5000, 500, 50, 5 .... **(1 mark)**

3 10^3, 10^2, 10 .... **(1 mark)**

4 The nth term of a geometric sequence is given by the formula $u_n = 2 \times 3^{n-1}$.
 (a) Calculate u_1, u_2 and u_3.

 ....

 ....

 .... **(3 marks)**

 (b) What is the value of n if $u_n = 1458$?

 ....

 .... **(2 marks)**

5 Part of a geometric sequence is given below:

 ..., ..., 4 , ..., ..., $\dfrac{1}{16}$ where $u_3 = 4$ and $u_6 = \dfrac{1}{16}$

Calculate:
 (a) the common ratio r

 ....

 .... **(2 marks)**

 (b) the value of u_1

 ....

 .... **(2 marks)**

 (c) the formula for the nth term

 ....

 .... **(2 marks)**

 (d) the 10th term, giving your answer as a fraction.

 ....

 .... **(2 marks)**

(16) Variation

○ Exercise 16.1

1 (a) If d is proportional to p and the constant of proportionality is k, write an equation for d in terms of p.

....(1 mark)

(b) If $d = 10$ when $p = 5$ find k. ....(1 mark)

(c) Find d when $p = 20$. ....(1 mark)

(d) Find p when $d = 2$. ....(1 mark)

2 a is inversely proportional to b.

(a) If k is the constant of proportionality, write an equation for a in terms of b.

....(1 mark)

(b) If $k = 20$, find a when $b = 40$. ....(1 mark)

3 p is inversely proportional to q squared. If $q = 0.5$ when $p = 2$:

(a) Write an equation for p in terms of q. ....(1 mark)

(b) Find p when $q = 5$. ....(1 mark)

(c) If $p = 0.005$ find two values for q.

....

....(2 marks)

4 q is proportional to p squared and q is inversely proportional to r cubed.
Using k as the final constant of proportionality, write an equation for p in terms of r.

....

....

....

....(3 marks)

○ Exercise 16.2

1 a is proportional to the cube of b. If $b = 2$ when $a = 32$, find a when $b = 5$.

....

....(2 marks)

○ Exercise 16.3

1 The power of an engine is proportional to the square of its mass. If an engine weighing 10 kg gives 200 b.h.p. find:

(a) the power of an engine weighing 30 kg

....

.... (2 marks)

(b) the mass of an engine giving 5000 b.h.p.

....

.... (2 marks)

2 The speed (v) in metres/seconds of a dam outlet is measured. It is proportional to the square root of the level indicated on a gauge (l) in metres. If l = 64 when v = 24, calculate v when l = 12 100.

....

....

.... (3 marks)

3 The force (f) Newtons between two objects is inversely proportional to the square of the distance (l) metres between them. Two magnets attract with a force of 18 Newtons when they are 2 cm apart.
What is the force of attraction when they are 6 cm apart?

....

....

.... (3 marks)

Graphs in practical situations

○ **Exercise 17.1**

1 Water is charged at $0.20 per unit.

   (a) Draw a conversion graph on the axes below up to 50 units.

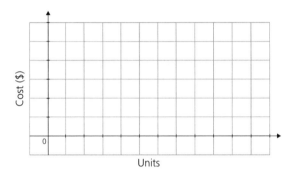

(3 marks)

   (b) From your graph, estimate the cost of using 23 units of water.

   Show your method clearly. .... (2 marks)

   (c) From your graph estimate the number of units used if the cost was $7.50

   Show your method clearly. .... (2 marks)

2 A Science exam is marked out of 180.

  (a) Draw a conversion graph to change the marks to percentages.

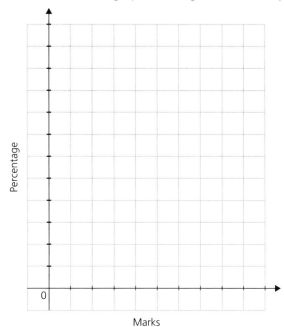

Percentage

0

Marks   **(3 marks)**

  (b) Using the graph and showing your method clearly, estimate the percentage score if a student achieved a mark of 130.

  .... **(2 marks)**

  (c) Using the graph and showing your method clearly, estimate the actual mark if a student got 35%.

  .... **(2 marks)**

◯ Exercise 17.2

1 Find the average speed of an object moving:

  (a) 60 m in 12 s ....**(1 mark)**

  (b) 140 km in 1 h 20 min ....**(2 marks)**

2 How far will an object travel during:

  (a) 25 s at 32 m/s ....**(1 mark)**

  (b) 2 h 18 min at 15 m/s? ....

  ....**(2 marks)**

3 How long will an object take to travel:

  (a) 2.5 km at 20 km/h ....**(1 mark)**

  (b) 4.8 km at 48 m/s? ....

  ....**(2 marks)**

○ **Exercises 17.3–17.4**

1 Two people, A and B, set off from points 300 m apart and travel towards each other along a straight road. A graph of their movement is shown below.

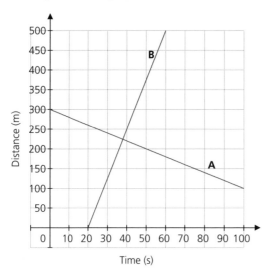

(a) Calculate the speed of person A.

.... (1 mark)

(b) Calculate the speed of person B when she is moving.

.... (1 mark)

(c) Use the graph to estimate how far apart they are 50 seconds after person A has set off.

.... (2 marks)

(d) Explain the motion of person B in the first 20 seconds.

.... (1 mark)

(e) Calculate the average speed of person B during the first 60 seconds.

.... (2 marks)

2   A cyclist sets off at 09 00 one morning and does the following:
   - Stage 1: Cycles for 30 minutes at a speed of 20 km/h
   - Stage 2: Rests for 15 minutes
   - Stage 3: Cycles again at a speed of 30 km/h for 30 minutes
   - Stage 4: Rests for another 15 minutes
   - Stage 5: Realises his front wheel has a puncture so walks with the bicycle for 30 minutes at a speed of 5 km/h to his destination.

   (a) At what time does the cyclist reach his destination?

   .... (1 mark)

   (b) How far does he travel during stage 1?

   .... (1 mark)

   (c) Draw a distance–time graph on the axes below, to show the cyclist's movement.
   Label all five stages clearly on the graph.

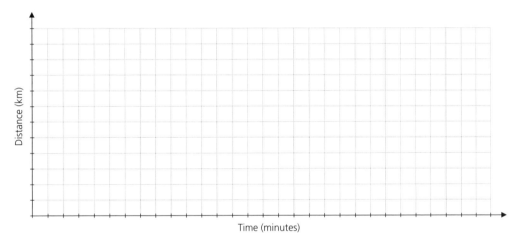

Time (minutes)

Distance (km)

   (5 marks)

   (d) Calculate the cyclist's average speed for the whole journey. Give your answer in km/h.

   .... (2 marks)

○ Exercises 17.5–17.6

1   Using the graphs below, calculate the acceleration/deceleration in each case.

(a)

(b)

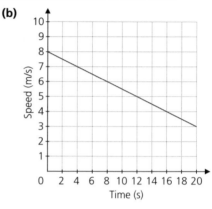

.... (1 mark)   ....(2 marks)

2   A sprinter is in training. Below is a graph of one of his sprints.

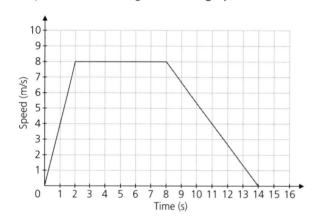

(a) Describe in words the sprinter's motion between the second and eighth second.

.... (1 mark)

(b) Calculate the acceleration/deceleration during the first two seconds.

....

.... (1 mark)

(c) Calculate the acceleration/deceleration during the last phase of the sprint.

....

.... (1 mark)

○ **Exercise 17.7**

1 A stone is dropped off the top of a cliff. It accelerates at a constant rate of $10\,\text{m/s}^2$ for 4 seconds before hitting the water at the bottom of the cliff.

(a) Complete the table below.

Time (s)	0	1	2	3	4
Speed (m/s)	0				

(2 marks)

(b) Plot a speed–time graph below for the 4 seconds it takes for the stone to drop.

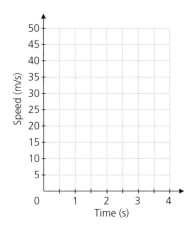

(2 marks)

(c) Calculate the distance fallen by the stone between the first and third seconds.

....

....

.... (2 marks)

(d) Calculate the height of the cliff.

....

.... (1 mark)

2 Two objects, A and B, are at the same point when the time $t = 0$ s.

At that point, object A accelerates from rest at a constant rate of $2 \, \text{m/s}^2$ for 6 seconds.

Object B is travelling at $15 \, \text{m/s}$ but decelerates at a constant rate of $3 \, \text{m/s}^2$ until it comes to rest.

(a) On the same axes below, plot a speed–time graph for both objects.

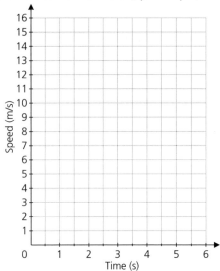

(4 marks)

(b) After how many seconds are the two objects travelling at the same speed?

....(1 mark)

(c) Assuming both objects are travelling in the same straight line, calculate how far apart they are after 4 seconds.

....

....

.... **(3 marks)**

Graphs of functions

○ **Exercise 18.1**

For each of the following quadratic functions, complete the table of values and draw the graph on the grid provided.

1 $y = 2x^2 + 12x + 16$

x	−5	−4	−3	−2	−1
y					

(2 marks)

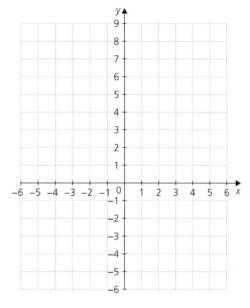

(2 marks)

2 $y = -x^2 + 3x + 4$

x	−2	−1	0	1	2	3	4	5
y								

(2 marks)

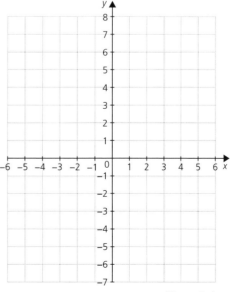

(2 marks)

○ **Exercise 18.2**

Solve each of the following quadratic functions below by first plotting a graph of the function.

1 $2x^2 - 8x - 10 = 0$

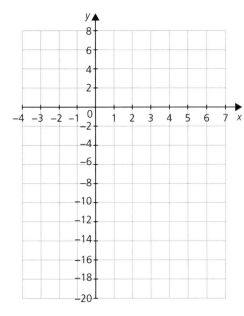

x = ...(2 marks)

(3 marks)

2 $-2x^2 + 16x - 24 = 0$

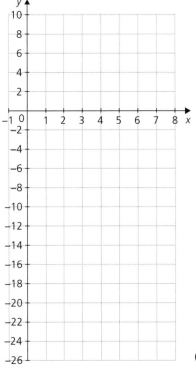

x = ...(2 marks)

(3 marks)

3 $-\dfrac{1}{2}x^2 - x + 24 = 0$

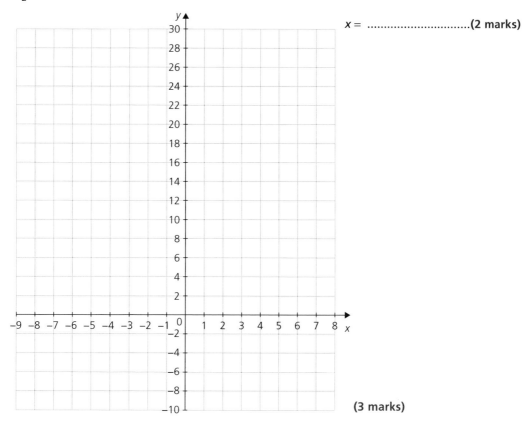

$x = $ ....(2 marks)

(3 marks)

○ **Exercise 18.3**

Using the graphs you drew in the previous exercise, solve the following quadratic equations. Show your method clearly.

1 $2x^2 - 8x + 6 = 0$

....(2 marks)

2 $-2x^2 + 16x - 30 = 0$

....(2 marks)

3 $-\dfrac{1}{2}x^2 - x + 12 = 0$

....(2 marks)

○ **Exercise 18.4**

1 Complete the table of values and draw on the grid provided the graph of the reciprocal function $y = \dfrac{3}{2x}$.

x	−4	−3	−2	−1	0	1	2	3	4
y									

(2 marks)

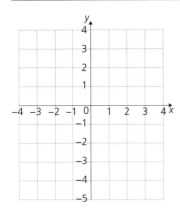

(2 marks)

○ **Exercises 18.5–18.6**

For each of the functions given below:
(a) draw up a table of values for x and f(x)
(b) plot a graph of the function.

1 $f(x) = \dfrac{1}{x^2} - x$ $-4 \leqslant x \leqslant 3$

(a)   **(b)**

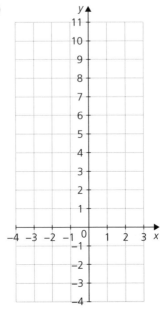

(2 marks)   (3 marks)

2 $f(x) = 3^x - x - 2$ $-5 \leqslant x \leqslant 2$

(a)

(b)

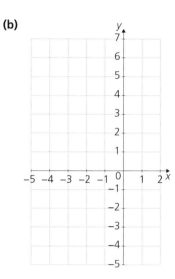

(2 marks)   **(3 marks)**

○ **Exercise 18.7**

For each of the functions below:

(a) plot a graph

(b) calculate the gradient of the function at the point given.

1 $y = x^2 - x - 2$ $-2 \leqslant x \leqslant 3$ Gradient where $x = 2$

(a)

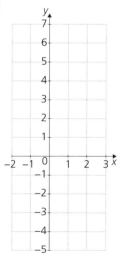

(2 marks)

(b) ....

....

.... **(3 marks)**

2 $y = 2x^{-1} + x$ $1 \leqslant x \leqslant 6$ Gradient where $x = 2$

(a)

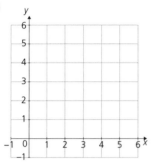

(2 marks)

(b) ....

....

.... (3 marks)

○ **Exercise 18.8**

1 (a) Plot the function $y = \dfrac{3}{x^2} - 2x$ for $-5 \leqslant x \leqslant 2$.

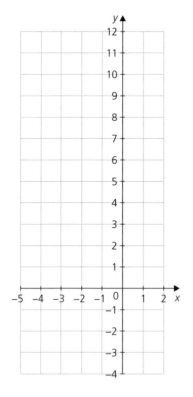

(3 marks)

(b) Showing your method clearly, use the graph to solve the equation
$x^3 + 4x^2 - 3 = 0$.

....

....

....

.... **(4 marks)**

2 **(a)** Plot the function $y = 3^x + \dfrac{1}{2}x$ for $-4 \leqslant x \leqslant 2$.

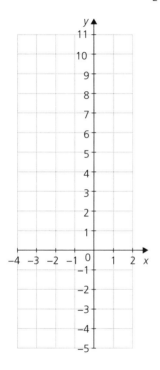

(3 marks)

(b) Showing your method clearly, use the graph to solve the equation
$2(3^x) - 3x - 8 = 0$.

....

....

....

.... **(4 marks)**

(19) Functions

○ Exercise 19.1

1 If $f(x) = 3x + 3$, calculate:

(a) $f(2)$ .... (1 mark)

(b) $f(4)$ .... (1 mark)

(c) $f\left(\dfrac{1}{2}\right)$ .... (1 mark)

(d) $f(-2)$ .... (1 mark)

(e) $f(-6)$ .... (1 mark)

(f) $f\left(-\dfrac{1}{2}\right)$ .... (1 mark)

2 If $f(x) = 2x - 5$, calculate:

(a) $f(4)$ .... (1 mark)

(b) $f(7)$ .... (1 mark)

(c) $f\left(\dfrac{7}{2}\right)$ .... (1 mark)

(d) $f(-4.25)$ .... (1 mark)

3 If $g(x) = -x + 6$, calculate:

(a) $g(0)$ .... (1 mark)

(b) $g(4.5)$ .... (1 mark)

(c) $g(-6.5)$ .... (1 mark)

(d) $g(-2.3)$ .... (1 mark)

○ Exercise 19.2

1 If $f(x) = \dfrac{2x}{3} + 4$ calculate:

(a) $f(3)$ ....

.... (2 marks)

(b) $f(9)$ ....

.... (2 marks)

(c) $f(-0.9)$ ....

.... (2 marks)

(d) $f(-1.2)$ ....

.... (2 marks)

2 If $g(x) = \dfrac{7x}{2} - 3$, calculate:

(a) g(2) ....

.... (2 marks)

(b) g(0) ....

.... (2 marks)

(c) g(–4) ....

.... (2 marks)

(d) g(–0.2) ....

.... (2 marks)

3 If $h(x) = \dfrac{-18x}{4} + 2$, calculate:

(a) h(1) ....

.... (2 marks)

(b) h(6) ....

.... (2 marks)

(c) h(–4) ....

.... (2 marks)

(d) h(–0.8) ....

.... (2 marks)

○ **Exercise 19.3**

1 If $f(x) = x^2 + 7$, calculate:

(a) f(11)...

.... (2 marks)

(b) f(1.1)...

.... (2 marks)

(c) f(–13) ....

.... (2 marks)

(d) $f\left(\dfrac{1}{2}\right)$ ....

.... (2 marks)

(e) $f\left(\sqrt{2}\right)$ ....

.... (2 marks)

→

2 If $f(x) = 2x^2 - 1$, calculate:

(a) $f(5)$ ....

.... **(2 marks)**

(b) $f(-12)$ ....

.... **(2 marks)**

(c) $f\left(\sqrt{3}\right)$ ....

.... **(2 marks)**

(d) $f\left(-\dfrac{1}{3}\right)$ ....

.... **(2 marks)**

3 If $g(x) = -5x^2 + 1$, calculate:

(a) $g\left(\dfrac{1}{2}\right)$ ....

.... **(2 marks)**

(b) $g(-4)$ ....

.... **(2 marks)**

(c) $g\left(\sqrt{5}\right)$ ....

.... **(2 marks)**

(d) $g\left(-\dfrac{3}{2}\right)$ ....

.... **(2 marks)**

○ Exercise 19.4

1 If $f(x) = 3x + 1$, write down the following in their simplest form:

(a) $f(x + 2)$ ....

.... **(3 marks)**

(b) $f(2x - 1)$ ....

.... **(3 marks)**

(c) $f(2x^2)$ ....

.... **(3 marks)**

(d) $f\left(\dfrac{x}{2} + 2\right)$ ....

.... **(3 marks)**

2 If $g(x) = 2x^2 - 1$, write down the following in their simplest form:

(a) $g(3x)$ ....

....

.... **(3 marks)**

(b) $g\left(\dfrac{x}{4}\right)$ ....

....

.... **(3 marks)**

(c) $g\left(\sqrt{2x}\right)$ ....

....

.... **(3 marks)**

(d) $g(x - 5)$ ....

....

.... **(3 marks)**

○ Exercise 19.5

Find the inverse of each of the following functions.

1 (a) $f(x) = x + 4$ ....

.... **(2 marks)**

(b) $f(x) = 5x$ ....

.... **(2 marks)**

2 (a) $g(x) = 3x - 5$ ....

....

.... **(3 marks)**

(b) $g(x) = \dfrac{5x}{2} - 1$ ....

....

.... **(3 marks)**

(c) $g(x) = \dfrac{2(2x - 3)}{5}$ ....

....

.... **(3 marks)**

○ **Exercise 19.6**

1 If $f(x) = x - 1$, evaluate:

(a) $f^{-1}(2)$ ....

....

.... (3 marks)

(b) $f^{-1}(0)$ ....

.... (1 mark)

2 If $f(x) = 2x + 3$ evaluate:

(a) $f^{-1}(5)$ ....

....

.... (3 marks)

(b) $f^{-1}(-1)$ ....

.... (1 mark)

3 If $g(x) = 3(x - 2)$, evaluate $g^{-1}(12)$. ....

....

.... (3 marks)

4 If $g(x) = \dfrac{x}{2} + 1$, evaluate $g^{-1}\left(\dfrac{1}{2}\right)$. ....

....

.... (3 marks)

○ **Exercise 19.7**

1 Write a formula for $fg(x)$ in each of the following:

(a) $f(x) = 2x$, $g(x) = x + 4$ ....

....

....

.... (3 marks)

(b) $f(x) = x + 4$, $g(x) = x - 4$ ....

....

....

.... (3 marks)

2 Write a formula for pq(x) in each of the following:

(a) p(x) = 2x, q(x) = x + 1 ....

....

.... **(3 marks)**

(b) p(x) = x + 1, q(x) = 2x ....

....

.... **(3 marks)**

3 Write a formula for jk(x) in each of the following:

(a) j(x) = $\dfrac{x-2}{4}$, k(x) = 2x ....

....

.... **(4 marks)**

(b) j(x) = 6x + 2, k(x) = $\dfrac{x-3}{2}$ ....

....

.... **(4 marks)**

4 Evaluate fg(2) in each of the following:

(a) f(x) = 3x − 2, g(x) = $\dfrac{x}{3}$ + 2 ....

....

.... **(4 marks)**

(b) f(x) = $\dfrac{2}{x+1}$, g(x) = −x + 1 ....

....

.... **(4 marks)**

○ **Exam focus**

1 **(a)** Factorise $9a^2 - 36b^2$.

....[2]

(b) Factorise $d^2 + d - 12$.

....[2]

(c) Make 'a' the subject of the formula $\dfrac{m}{n} = \dfrac{3c}{b} - \dfrac{a}{b}$.

....

....[3]

(d) Factorise and simplify $\dfrac{a^2 + 5a}{a^2 + 2a - 15}$.

....

....[3]

2 A cone has a height (h) of 8 cm and a volume (V) of 300 cm³. Given the formula $V = \dfrac{1}{3}\pi r^2 h$, find the radius r.

....

....

....[3]

3 Simplify the following:

(a) $4a^2 \times 3a^3$ ....[1]

(b) $2a^2b \times 4a^3b^2$ ....[1]

(c) $(a^2)^{\frac{1}{2}} \times \sqrt[3]{a^3}$ ....[2]

(d) $\dfrac{(a^{-2})^{-2}}{a^{-\frac{1}{2}} \times \left(\sqrt[4]{a}\right)^{-2}}$ ....

....[3]

4 Consider the shape below.

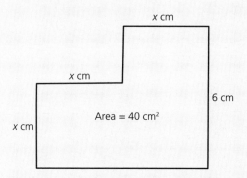

(a) Write an equation in terms of x for the area of the shape.

....

....[3]

(b) Solve the equation to find the value of x.

....

....[3]

5 (a) Write the equation $4x^2 - 24x + 29 = 0$ in completed square form.

....

....

....[3]

(b) Solve the equation $4x^2 - 24x + 29 = 0$ giving your answer(s) to 1 d.p.

....

....

....[4]

6 A man takes some chickens (x) and ducks (y) to sell at a market.
The number of chickens and twice the number of ducks is less than or equal to 30.
Three times the number of chickens and half the number of ducks is greater than 15.
The number of chickens is less than 4.

(a) Express the three conditions above as inequalities.

....

....

....[3]

(b) On the axes below, identify the region that satisfies all the inequalities by shading the **unwanted** regions.

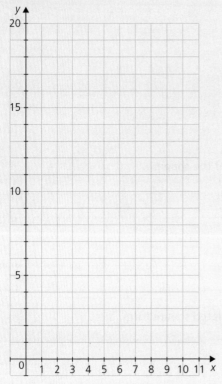

[6]

(c) State the possible solution(s) for the number of chickens and ducks that the man takes to market.

....[2]

7   For each of the sequences given below:
    (i)   calculate the next two terms
    (ii) calculate the rule for the nth term.

(a) 11, 22, 33, 44...[2]

....[2]

(b) 63, 56, 49, 42...[2]

....[2]

(c) 40, −20, 10, −5...[2]

....

....[3]

(d) 1, $\dfrac{1}{3}$, $\dfrac{1}{9}$, $\dfrac{1}{27}$ ....[2]

....

....[3]

8 *y* is inversely proportional to *x*. If *y* = 10 when *x* = 4, find *y* when *x* = 5.

....

....**[2]**

9 An object is travelling as shown on the speed–time graph below. Its motion was recorded as four stages: A, B, C and D.

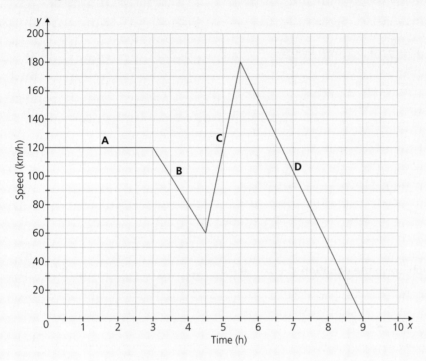

(a) During which stage(s) was the object travelling at constant speed? Justify your answer.

....

....**[2]**

(b) What was the rate of acceleration during stage C?

....

....**[2]**

(c) During which stage was the deceleration greatest? Justify your answer by calculation and by referring to the shape of the graph.

....

....

....**[2]**

→

(d) What is the distance travelled during stage C?

....

....[2]

(e) What is the total distance travelled by the object?

....

....[3]

10 The function $y = \dfrac{4}{x^2} - 3$ is shown below.

Using the graph, solve the equation $x^3 + 5x^2 - 4 = 0$.

....

....

....[4]

11 If $h(x) = \dfrac{1}{3}x - 3$, evaluate $h^{-1}\left(-\dfrac{1}{2}\right)$.

....

....

....[4]

12 The functions g(x) and h(x) are given below:

$g(x) = 10(3x - 1)$, $h(x) = \dfrac{2x}{5}$

(a) Write a formula for gh(x).

....

....

....[3]

(b) Evaluate gh(−4).

....

....[2]

20 Geometrical vocabulary

○ **Exercise 20.1**

1 Explain, giving reasons, whether the following triangles are **definitely** congruent.
(Note: the diagrams are not drawn to scale.)

(a)

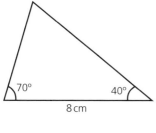

.... **(2 marks)**

(b)

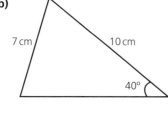 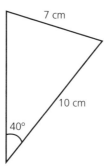

.... **(2 marks)**

○ **Exercise 20.2**

1 Complete the table below, by entering either 'Yes' or 'No' in each cell.

	Rhombus	Parallelogram	Kite
Opposite sides equal in length			
All sides equal in length			
All angles right angles			
Both pairs of opposite sides parallel			
Diagonals equal in length			
Diagonals intersect at right angles			
All angles equal			

(3 marks)

○ **Exercise 20.3**

1 Three nets A, B and C are shown below. Which (if any) can be folded to make a cube?

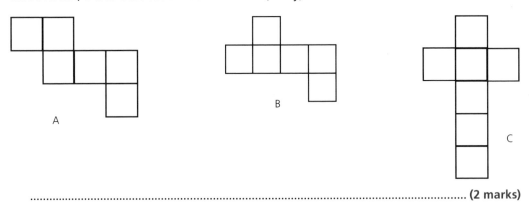

A

B

C

.... **(2 marks)**

Geometrical constructions and scale drawings

○ **Exercises 21.1–21.3**

1 Using only a ruler and a pair of compasses, construct the following triangle XYZ.
XY = 5 cm, XZ = 3 cm and YZ = 7 cm.

(3 marks)

2 Draw an angle of 300° below. Using a pair of compasses, bisect the angle.

(3 marks)

→

3 On the triangle ABC below:
   (a) Construct the perpendicular bisector of each of the triangle's sides.   **(3 marks)**
   (b) Draw a circle such that the circumference passes through each of the vertices
        A, B and C.   **(1 mark)**

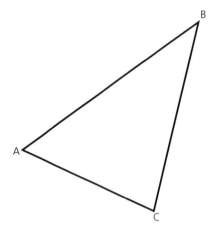

○ **Exercise 21.4**

1 The scale of a map is 1 : 20 000.
   (a) Two villages are 12 cm apart on the map. How far apart are they in real life?
        Give your answer in kilometres.
        .... (1 mark)
   (b) The distance from a village to the edge of a lake is 8 km in real life.
        How far apart would they be on the map? Give your answer in centimetres.
        .... (1 mark)

2 **(a)** A model car is a $\frac{1}{25}$ scale model. Express this as a ratio.
        .... (1 mark)
   (b) If the length of the real car is 4.9 m, what is the length of the model car?
        .... (1 mark)

(22) Similarity

○ **Exercise 22.1**

1 Two triangles are shown below.

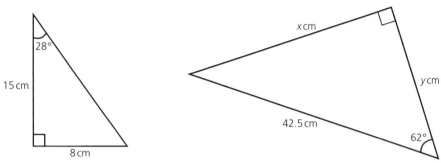

(a) Explain why the two triangles are similar.

....

.... (2 marks)

(b) Calculate the scale factor that enlarges the smaller triangle to the larger triangle.

....

.... (3 marks)

(c) Calculate the value of *x*.

.... (1 mark)

(d) Calculate the value of *y*.

.... (1 mark)

2

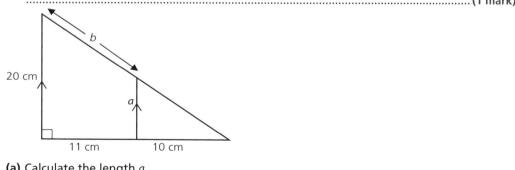

(a) Calculate the length *a*.

....

.... (2 marks)

(b) Calculate the length *b*.

....

....

.... (3 marks)

○ **Exercise 22.2**

1 The five rectangles below are each an enlargement of the previous one by a scale factor of 1.2.

(a) If the area of rectangle D is 100 cm², calculate to 1 d.p. the area of:

  (i) rectangle E ....(1 mark)

  (ii) rectangle A. ....

  ....(2 marks)

(b) If the rectangles were to continue in this sequence, which letter rectangle would be the last to have an area below 500 cm²? Show your method clearly.

  ....

  ....

  ....

  ....(3 marks)

2 A triangle has an area of 50 cm². If the lengths of its sides are all reduced by a scale factor of 30%, calculate the area of the reduced triangle.

  ....

  ....

  ....(3 marks)

○ **Exercises 22.3–22.4**

1 A cube has a side length of 4.5 cm.

(a) Calculate its total surface area.

  ....

  ....(2 marks)

(b) The cube is enlarged and has a total surface area of 1093.5 cm².
    Calculate the scale factor of enlargement.

  ....

  ....(3 marks)

(c) Calculate the volume of the enlarged cube.

  ....

  ....(2 marks)

2   The two cylinders shown below are similar.

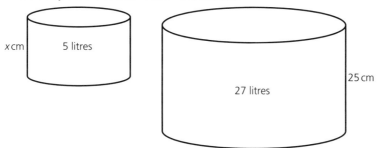

(a) Calculate the volume factor of enlargement.

.... (1 mark)

(b) Calculate the scale factor of enlargement. Give your answer to 2 d.p.

....

.... (2 marks)

(c) Calculate the value of x.

.... (1 mark)

3   A large cone has its top sliced as shown in the diagram below.
    The smaller cone is mathematically similar to the original cone.

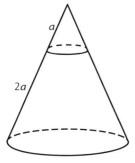

(a) What is the scale factor of enlargement from the small cone to the original cone?

.... (1 mark)

(b) If the original cone has a volume of 1350 cm³, calculate the volume of the smaller cone.

....

.... (3 marks)

4   An architect's drawing is drawn to a scale of 1:50. The area of a garden on his drawing is
    620 cm². Calculate the area of the real garden, giving your answer in m².

....

....

.... (4 marks)

(23) Symmetry

○ Exercise 23.1

1   On each of the pairs of diagrams below, draw a different plane of symmetry.

(a) A cuboid with a square cross-section.

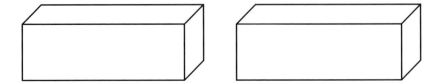

(2 marks)

(b) A triangular prism with an isosceles triangular cross-section.

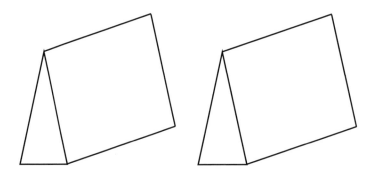

(2 marks)

2   Determine the order of rotational symmetry of the cube, about the axis given.

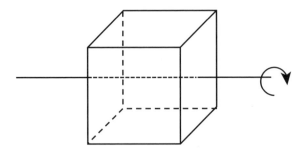

.... (2 marks)

○ **Exercise 23.2**

1 In the circle below, O is the centre, AB = CD and X and Y are the midpoints of AB and CD respectively. Angle OCD = 50° and angle AOD = 30°.

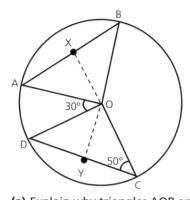

(a) Explain why triangles AOB and COD are congruent.

....

....

.... **(2 marks)**

(b) What type of triangle is triangle AOB? ....**(1 mark)**

(c) Calculate the obtuse angle XOY.

....

....

.... **(3 marks)**

○ **Exercise 23.3**

1 The diagram below shows a circle with centre at O. XZ and YZ are both tangents to the circle.

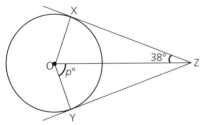

Calculate, giving detailed reasons, the size of the angle marked *p*.

....

....

....

....**(3 marks)**

(24) Angle properties

◯ Exercises 24.1–24.3

1   Calculate the size of each of the labelled angles below.

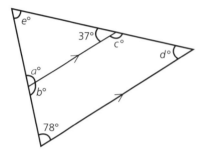

$a =$ ....(1 mark)

$b =$ ....(1 mark)

$c =$ ....(1 mark)

$d =$ ....(1 mark)

$e =$ ....(1 mark)

2   Calculate the size of the labelled angles in the kite below.

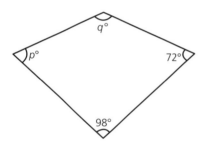

$p =$ ....(1 mark)

$q =$ ....(2 marks)

◯ Exercise 24.4

1   The size of each interior angle of a regular polygon is 165°. Calculate:
   (a) the size of each exterior angle

   ....(1 mark)

   (b) the number of sides of the regular polygon.

   .... (2 marks)

2   Find the value of each interior angle of a regular polygon with:

   (a) 30 sides ....

   .... (2 marks)

   (b) 20 sides. ....

   .... (2 marks)

3 Calculate the number of sides of a regular polygon if each interior angle is $4x°$ and each exterior angle $x°$.

....

....(3 marks)

○ **Exercise 24.5**

In each of the following diagrams, O marks the centre of the circle. Calculate the value of x in each case.

1

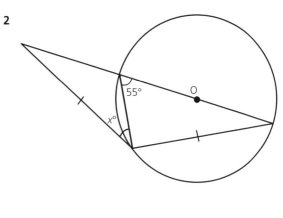

.... (2 marks)

2

.... (3 marks)

○ **Exercise 24.6**

In each of the following diagrams, O marks the centre of the circle. Calculate the value of x in each case.

1

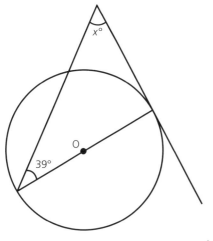

.... (2 marks)

2

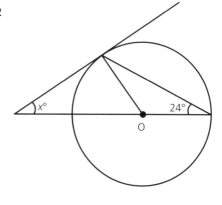

.... (3 marks)

○ **Exercise 24.7**

1 The pentagon below has angles as shown.

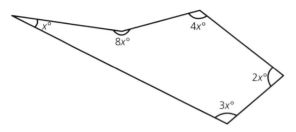

(a) State the sum of the internal angles of a pentagon.

.... (1 mark)

(b) Calculate the value of x.

....

.... (2 marks)

(c) Calculate the size of each of the angles of the hexagon.

....

.... (2 marks)

2 The diagram below shows an octagon.

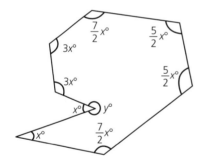

(a) Write the angle y in terms of x. .... (1 mark)

(b) Write an equation for the sum of the interior angles of the octagon in terms of x.

....

.... (2 marks)

(c) Calculate the value of x.

....

.... (2 marks)

(d) Calculate the size of the angle labelled y.

.... (1 mark)

○ **Exercise 24.8**

In each of the following diagrams, O marks the centre of the circle. Calculate the value of the marked angles in each case.

1

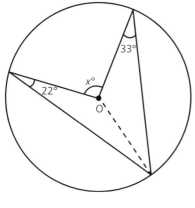

2

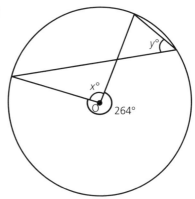

....(3 marks)   ....

....(3 marks)

○ **Exercise 24.9**

In the following, calculate the size of the marked angles.

1

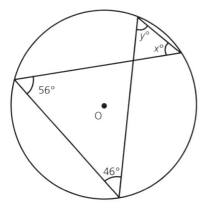

2

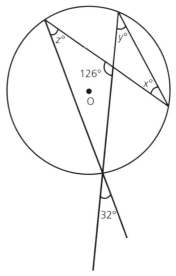

....

....(3 marks)   ....(3 marks)

○ **Exercise 24.10**

In the following, calculate the size of the marked angles.

1

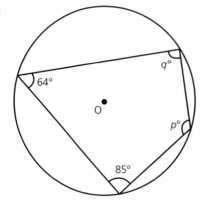

....
....(2 marks)

2

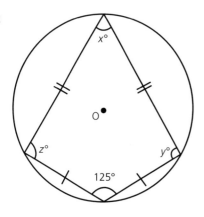

....
....(3 marks)

25 Loci

○ Exercises 25.1–25.3

1 The diagram below shows a plan view of a rail AB bent at 90° at C.
A horse is tethered to the rail by a ring attached to a rope 3 m long. The ring can run freely along the full length of the rail. The horse can access both sides of the rail.
Using a scale of 1 cm = 1 m, shade the locus of all the points that the horse can reach.

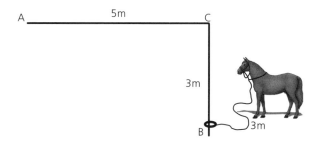

....(3 marks)

2 The diagram below is a plan view of a tall column and three people, one each at A, B and C. The people at A and B cannot see over the column. In the current position, the person at C cannot be seen by either of the people at A or B.

Assuming that A and B remain in the same place, shade the locus of the points where the person at C should stand if he wants to be seen by A but not B.   **(3 marks)**

○ **Exam focus**

1 Draw a possible net for the cuboid in the grid below.

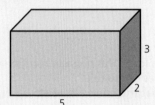

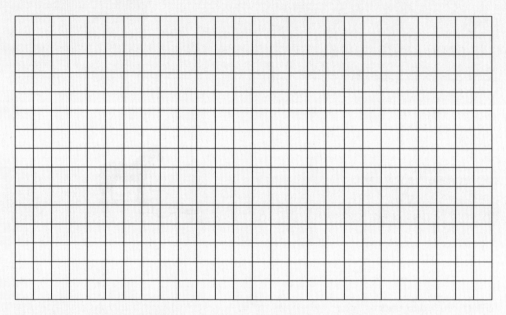

[3]

2 The scale of a map is 1:50 000.

(a) Two rivers are 8.2 cm apart on the map. How far apart are they in real life?
Give your answer in metres.

....[2]

(b) Two towns are 20 km apart in real life. How far apart are they on the map?
Give your answer in centimetres.

....[2]

3 A cube is enlarged by increasing the lengths of its sides by 10%.

(a) Calculate the volume factor of enlargement from the original to the enlarged cube.

....[2]

(b) If the volume of the enlarged cube is 2662 cm^3, calculate the volume of the original cube.

....[2]

4   Two right-angled triangular prisms are shown below. One is an enlargement of the other.
    (Knowledge of Pythagoras' theorem is needed for this question.)

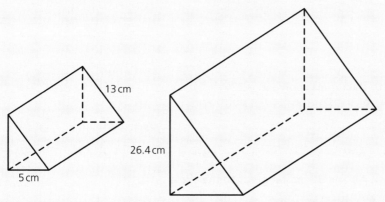

13 cm

26.4 cm

5 cm

(a) Calculate the scale factor of enlargement.

....

....[3]

(b) Calculate the area factor of enlargement.

....[1]

(c) Calculate the volume factor of enlargement.

....[1]

(d) If the volume of the larger prism is 5000 cm³, calculate the volume of the smaller prism.

....

....[2]

5   Calculate the angle marked x in the diagram below. O is the centre of the circle and PQ and
    QR are both tangents to the circle.

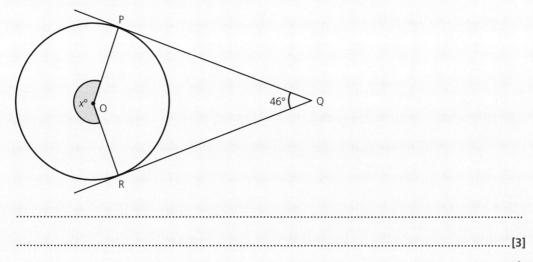

....

....[3]

→

6 The diagram below shows only two of the sides of a regular polygon.

176°

Each internal angle of the regular polygon is 176°.
Calculate the number of sides of the regular polygon.

....

....**[3]**

7 In the diagram below, AB and BC are tangents to the circle.
AB = 8 cm, OB = x cm and angle OAC = 24°.

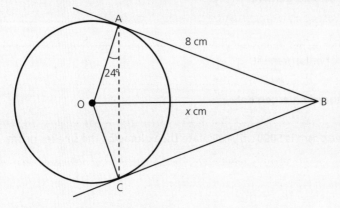

A

8 cm

24°

O

x cm

B

C

(a) Calculate the angle ABO. ....

....**[2]**

(b) Calculate the value of x.

....

....

....**[3]**

8 The circle below has its centre at O.

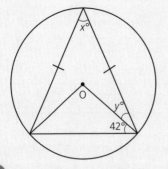

x°

O

y°

42°

(a) Calculate the size of the angle *x*.

....

....[3]

(b) Calculate the size of the angle *y*.

....

....[2]

9 A trapezium is shown in the circle below. Each of its vertices lies on the circumference of the circle.

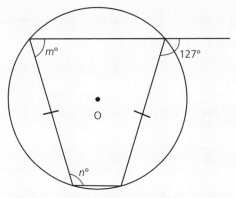

(a) Calculate the size of angle *n*.

....[2]

(b) Calculate the size of angle *m*.

....[2]

10 A rectangular garden ABCD of dimensions 8 m × 4 m is shown below.

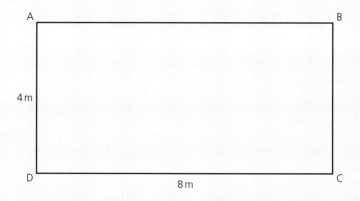

A landscape designer wants to construct a path through the garden so that its central line is always equidistant from the corners A and C. If the path is 1 m wide, construct and shade the locus of the points representing the path. [4]

Mensuration

(26) Measures

○ **Exercises 26.1–26.5**

1 Convert the following lengths to the units indicated:

(a) 0.072 m to mm .... (1 mark)

(b) 20 400 m to km .... (1 mark)

2 Convert the following masses into the units indicated:

(a) 420 g to kg .... (1 mark)

(b) 1.04 tonnes to kg .... (1 mark)

3 Convert the following liquid measures into the units indicated:

(a) 12 ml to litres .... (1 mark)

(b) 0.24 litres to ml .... (1 mark)

4 A rectangular field has an area of 105 000 m². Convert the area into km².

....

.... (2 marks)

5 **(a)** A container has a volume of 3.6 m³. Convert the volume into cm³.

.... (2 marks)

(b) A box has a volume of 3250 cm³. Convert the volume into:

(i) mm³ .... (2 marks)

(ii) m³ .... (2 marks)

Perimeter, area and volume

○ **Exercises 27.1–27.5**

1 Calculate the circumference and area of the circle below.

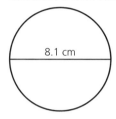

8.1 cm

....
....**(2 marks)**

2 A circle has an area of 12.25π cm². Calculate:

(a) its diameter.. **(2 marks)**

(b) its circumference. .... **(2 marks)**

3 A semicircular shape is removed from a trapezium shape as shown.

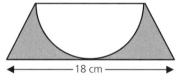

—— 18 cm ——

If the semicircle has a radius of 7 cm, calculate the shaded area remaining.

....
....**(3 marks)**

4 A trapezium and parallelogram are joined as shown.

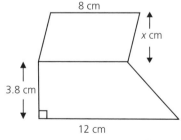

8 cm

x cm

3.8 cm

12 cm

If the total area is 53.2 cm², calculate the value of x.

....
....**(3 marks)**

→

5 Five thin semicircular chocolate pieces are placed in a rectangular box as shown.

Calculate:

(a) the length of the box .... (1 mark)

(b) the area occupied by one piece of chocolate

.... (1 mark)

(c) the area of the box not covered by chocolate pieces.

....

.... (2 marks)

6 A circular hole is cut out of a circular shape as shown.
The area remaining is the same as the area of the hole removed. The circumference of the original piece is 15π cm.

Calculate the radius of the circular hole.

....

....

.... (4 marks)

○ **Exercises 27.6–27.9**

1 A cuboid has a length of 7 cm, a width of 2.5 cm and a total surface area of 114.8 cm^2.
Calculate its height.

....

.... (2 marks)

2 A cylinder has a total surface area of 100 cm^2.
If the radius of its circular cross-section is 3.6 cm, calculate its height.

....

....

.... (2 marks)

3 A cylinder and cuboid have dimensions as shown.

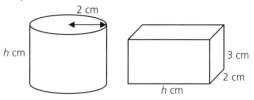

(a) Write an expression for the total surface area of the cuboid.

.... (2 marks)

(b) Write an expression for the total surface area of the cylinder.

.... (2 marks)

(c) If the total surface area of the cylinder is twice that of the cuboid, find the value of h.

....

....

.... (3 marks)

4 A metal hand weight is made from two cubes and a cylinder joined as shown:

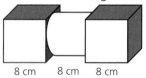

8 cm   8 cm   8 cm

Calculate the total volume of the shape.

....

.... (3 marks)

5 A cylinder and a cylindrical pipe have the same volume and diameter.

12 cm   30 cm

(a) Calculate the volume of the solid cylinder.

.... (2 marks)

(b) Write an expression for the volume of the cylindrical pipe.

....

....

.... (3 marks)

(c) Calculate the value of x.

....

....

.... (3 marks)

○ Exercises 27.10–27.11

1 A sector has a radius of 6.7 cm and an arc length of 3.2 cm. Calculate the angle of the sector θ.

....

....(1 mark)

2 A sector has an angle of 162° and an arc length of 14.5 cm. Calculate the length of the sector's radius.

....

....(1 mark)

3 The diagram below shows two arcs with the same centre and an angle θ.

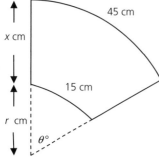

(a) Write an expression for the length x in terms of r.

.... (1 mark)

(b) If the perimeter of the shape is 100 cm, calculate the size of θ.

....

....

.... (3 marks)

○ Exercises 27.12–27.13

1 Two sectors A and B are shown below.

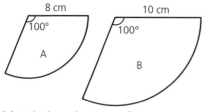

(a) Calculate the area of sector A.

....(1 mark)

(b) What is the ratio of the areas of sectors A : B?
Give your answer in the form 1 : n.

.... (2 marks)

2   A prism with a cross-section in the shape of a sector is shown below.

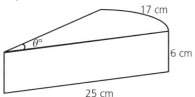

Calculate:

(a) the angle θ

.... (1 mark)

(b) the area of the cross-section

....

.... (2 marks)

(c) the total surface area of the prism

....

.... (3 marks)

(d) the volume of the prism.

.... (1 mark)

○ **Exercises 27.14–27.15**

1   A sphere has a volume of $0.5\,m^3$. Calculate the sphere's radius, giving your answer in cm to 1 d.p.

....

.... (2 marks)

2   A hemispherical bowl, with an outer radius of 20 cm, is shown below. A sphere is placed inside the bowl. The size of the sphere is such that it just fits the inside of the bowl.

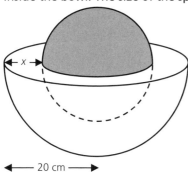

(a) Explain why the expression for the volume of the hemispherical bowl in terms of x can be written as $\frac{2}{3}\pi \times 20^3 - \frac{2}{3}\pi(20-x)^3$.

....

.... (2 marks)

→

(b) Write an expression for the volume of the sphere in terms of x.

.... (2 marks)

(c) If both the bowl and sphere have the same volume, show that $3(20 - x)^3 = 8000$.

....

.... (2 marks)

(d) Calculate the thickness x of the bowl.

....

.... (2 marks)

○ Exercise 27.16

1 A hemisphere has a radius of 5 cm.

(a) Calculate the area of its base.

.... (1 mark)

(b) Calculate its total surface area.

....

.... (3 marks)

2 A solid shape is made from two hemispheres joined together. The base of each hemisphere shares the same centre.

(a) Calculate the surface area of the smaller hemisphere.

.... (2 marks)

(b) Calculate the total surface area of the shape.

....

....

....

.... (4 marks)

○ **Exercises 27.17–27.19**

1 A rectangular-based pyramid is shown below.

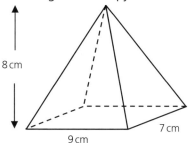

8 cm

7 cm

9 cm

Calculate:

(a) the volume of the pyramid

.... **(2 marks)**

(b) the total surface area of the pyramid, using Pythagoras' theorem.

....

....

.... **(4 marks)**

2 Two square-based pyramids are joined at their bases. The bases have an edge length of 6 cm.

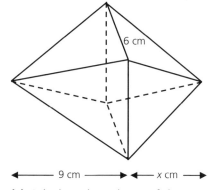

6 cm

← 9 cm → ← x cm →

(a) Calculate the volume of the pyramid on the left.

.... **(2 marks)**

(b) If the volume of the pyramid on the left is twice that of the pyramid on the right, calculate the value of x.

....

.... **(2 marks)**

(c) By using Pythagoras as part of the calculation, calculate the total surface area of the shape.

....

....

.... **(4 marks)**

⃝ **Exercises 27.20–27.23**

1 A cone has a base diameter of 8 cm and a sloping face length of 5 cm.

(a) Calculate its perpendicular height.

.... (2 marks)

(b) Calculate the volume of the cone.

....

.... (2 marks)

(c) Calculate the total surface area of the cone.

....

....

.... (4 marks)

2 Two similar sectors are shown below.

(a) Calculate the length of the radius *r*.

....

.... (2 marks)

(b) What is the value of *R*?

.... (1 mark)

The sectors are assembled to form cones.

(c) Calculate the volume of the smaller cone.

....

....

....

.... (4 marks)

(d) Calculate the curved surface area of the large cone.

....

.... (2 marks)

3 A cone of base radius 10 cm and a vertical height of 20 cm, has a cone of base radius 10 cm and a vertical height 10 cm removed from its inside as shown below.

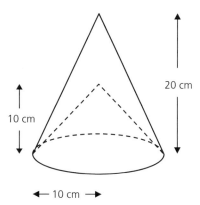

20 cm

10 cm

← 10 cm →

(a) Calculate the volume of the small cone removed from the inside.

.... **(2 marks)**

(b) Calculate the volume of the shape that is left (i.e the volume of the large cone with the small cone removed).

....

.... **(2 marks)**

(c) Calculate the total curved surface area of the final shape.

....

....

....

....

.... **(5 marks)**

○ **Exam focus**

1   A field has an area of $8700\,m^2$. Convert the area into km^2.

....[2]

2   A swimming pool has a volume of $800\,m^3$. Give the volume in cm^3 written in standard index form.

....[3]

3   Two cylinders A and B are shown below. The total surface area of B is twice that of A.

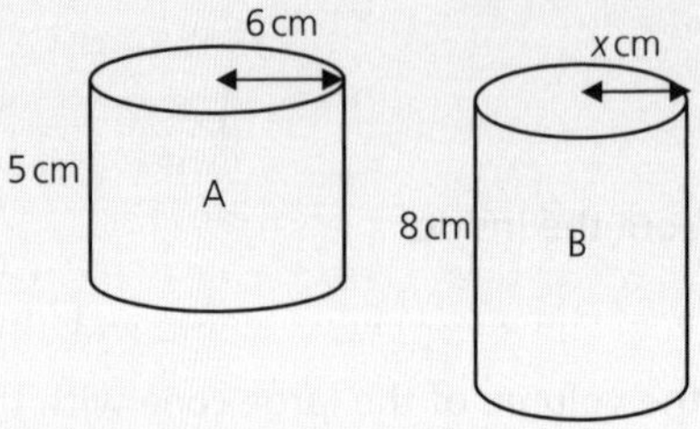

6 cm

5 cm

A

x cm

8 cm

B

Calculate:

(a) the surface area of cylinder A

....

....[2]

(b) the value of x

....

....

....[3]

(c) the ratio of the volume A : B in the form 1 : n.

....

....

....[3]

4 A shape is made from two sectors arranged in such a way that they share the same centre.
The radius of the smaller sector is 6 cm, whilst the radius of the larger sector is 10 cm.
The angle at the centre of the smaller sector is 22° as shown.

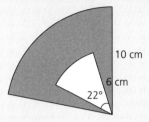

(a) Calculate the area of the smaller sector.

....[2]

(b) If the shaded area is four times the area of the smaller sector, calculate the angle at the
centre of the large sector.

....

....[3]

5 Two identical spheres are placed inside a box, so that they just fit.
The length of the box is 30 cm as shown.

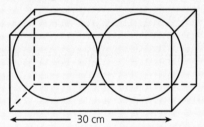

(a) Calculate the volume of one of the spheres.

....[2]

(b) Calculate the percentage volume of the box that is occupied by the spheres.

....

....[2]

6 The solid shape below is made up of a cone, a cylinder and a hemisphere.
The length of each of the three pieces is 5 cm.

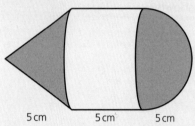

5 cm   5 cm   5 cm

(a) What is the diameter of the base of the cone? ....[1]

(b) Calculate:

    (i) the volume of the cone

    ....

    ....[2]

    (ii) the total volume of the shape.

    ....

    ....

    ....[3]

(c) Calculate:

    (i) the surface area of the cone

    ....

    ....

    ....[3]

    (ii) the total surface area of the shape.

    ....

    ....

    ....[3]

Coordinate geometry

 Straight-line graphs

○ **Exercises 28.1–28.3**

In each of the following, identify the coordinates of some of the points on the line and use these to find

   (a) the gradient of the line

   (b) the equation of the straight line.

1

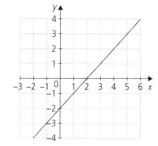

.... (2 marks)

2

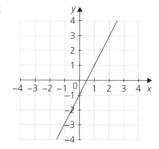

.... (2 marks)

3

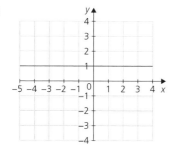

.... (2 marks)

→

4

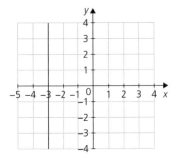

....**(2 marks)**

5

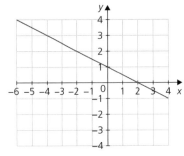

....**(2 marks)**

In each of the following, identify the coordinates of some of the points on the line and use these to find the equation of the straight line.

6

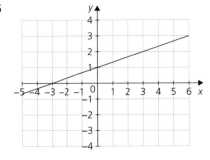

....**(3 marks)**

7

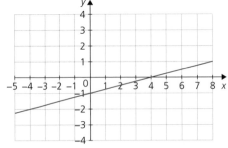

....**(3 marks)**

8

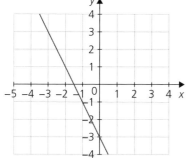

.... (3 marks)

9

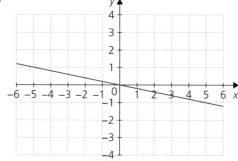

.... (3 marks)

10

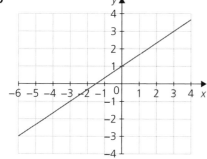

.... (4 marks)

11 The general equation of a straight line takes the form $y = mx + c$. In your own words, explain the significance of 'm' and 'c'.

....

.... (3 marks)

○ **Exercise 28.4**

For the following linear equations, calculate both the gradient and *y*-intercept in each case.

1 **(a)** $y = 4x - 2$

....

....

.... **(2 marks)**

(b) $y = -(2x + 6)$

....

....

..................................**(2 marks)**

2 **(a)** $y + \frac{1}{2}x = 3$

....

....

....

.... **(3 marks)**

(b) $y - (4 - 3x) = 0$

....

....

....

..................................**(3 marks)**

3 **(a)** $\frac{1}{2}y + x - 2 = 0$

....

....

....

....

.... **(4 marks)**

(b) $-5y - 1 - 10x = 0$

....

....

....

....

..................................**(4 marks)**

4 **(a)** $-\frac{4}{3}y + 2x = 4$

....

....

....

....

.... **(4 marks)**

(b) $\frac{3x - 2y}{5} = -3$

....

....

....

....

..................................**(4 marks)**

5 **(a)** $\frac{2y - x}{x - y} = 3$

....

....

....

....

.... **(4 marks)**

(b) $\frac{2x}{y + 1} + \frac{2}{3y + 3} = 2$

....

....

....

....

..................................**(4 marks)**

○ Exercise 28.5

1 Find the equation of the straight line parallel to $y = -2x + 6$ that passes through the point (2, 5).

....

....**(2 marks)**

2 Find the equation of the straight line parallel to $7y - 3x + 28 = 0$ that passes through the point $(7, \frac{11}{2})$. Give your answer in the form $ax + by + c = 0$.

....

....

....**(3 marks)**

○ Exercise 28.6

Plot the following straight lines.

1 $y = x - 3$

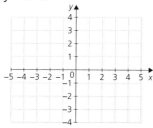

(2 marks)

2 $-y = 2x + 3$

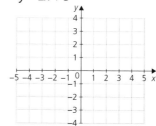

(2 marks)

3 $-3y + 6x - 4\frac{1}{2} = 0$

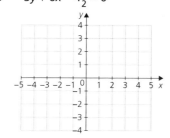

(3 marks)

4 $\dfrac{\frac{1}{2}x - y}{2} = 1$

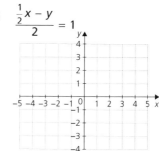

(3 marks)

5 $-\dfrac{x}{3} + \dfrac{y}{4} = \dfrac{1}{2}$

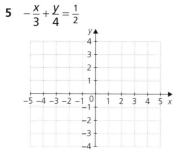

(4 marks)

○ **Exercise 28.7**

Solve the simultaneous equations below:

    (a) by graphical means
    (b) by algebraic means.

1 $y = \frac{1}{2}x + 4$ and $y + x + 2 = 0$

    (a)

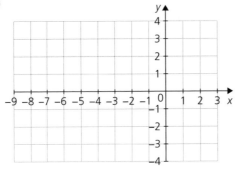

                                                                             (4 marks)

    (b) ....

         ....

         .... **(2 marks)**

2 $y + 3 = x$ and $3x + y - 1 = 0$

    (a)

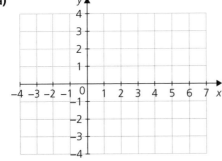

                                                                             (4 marks)

    (b) ....

         ....

         .... **(2 marks)**

○ **Exercise 28.8**

In each of the following:

    (a) calculate the length of the line segment between each of the pairs of points to 1 d.p.

    (b) calculate the coordinates of the midpoint of the line segment.

1 (7, 3) and (7, 9)

    (a) ....   **(b)** ....

    ....

    ....**(1 mark)**   ....**(1 mark)**

2 (3, 5) and (–2, 7)

    (a) ....   **(b)** ....

    ....

    ....

    ....**(2 marks)**   ....**(1 mark)**

3 (–2, –4) and (4, 0)

    (a) ....   **(b)** ....

    ....

    ....

    ....**(2 marks)**   ....**(1 mark)**

4 $(\frac{1}{2}, -3)$ and $(-\frac{1}{2}, 6)$

    (a) ....   **(b)** ....

    ....

    ....

    ....**(2 marks)**   ....**(2 marks)**

5 $\left(-\frac{1}{4}, -1\right)$ and $\left(-\frac{3}{4}, \frac{1}{2}\right)$

    (a) ....   **(b)** ....

    ....

    ....

    ....**(2 marks)**   ....**(2 marks)**

○ **Exercise 28.9**

Find the equation of the straight line which passes through each of the following pairs of points.

1 $(4, -2)$ and $(-6, -7)$

....

....**(2 marks)**

2 $(0, 6)$ and $\left(\frac{1}{2}, 5\right)$

....

....

....**(3 marks)**

3 $(-2, 7)$ and $(3, 7)$

....

....

....**(3 marks)**

4 $(-4, 2)$ and $\left(3, -\frac{3}{2}\right)$

....

....

....**(3 marks)**

5 $\left(\frac{1}{2}, -4\right)$ and $\left(\frac{1}{2}, -\frac{1}{2}\right)$

....

....

....**(3 marks)**

6 $(0, 4)$ and $(-5, 2)$

....

....

....**(3 marks)**

7 $\left(2, -\frac{14}{3}\right)$ and $\left(4, -\frac{16}{3}\right)$

....

....

....**(3 marks)**

○ **Exercise 28.10**

In each of the following, calculate:
(a) the gradient of the line joining the points
(b) the gradient of a line perpendicular to this line
(c) the equation of the perpendicular line if it passes through the first point each time.

1 (8, 3) and (10, 7)

(a) ....(1 mark)

(b) ....(1 mark)

(c) ....

....(2 marks)

2 (3, 5) and (4, 4)

(a) ....(1 mark)

(b) ....(1 mark)

(c) ....

....(2 marks)

3 (−3, −1) and (−1, 4)

(a) ....(1 mark)

(b) ....(1 mark)

(c) ....

....(2 marks)

4 (4, 8) and (−2, 8)

(a) ....(1 mark)

(b) ....(1 mark)

(c) ....

....(2 marks)

→

5 $\left(\frac{1}{2}, \frac{5}{2}\right)$ and $\left(3, -\frac{5}{4}\right)$

  (a) .... **(1 mark)**

  (b) .... **(1 mark)**

  (c) ....

  .... **(2 marks)**

6 $\left(-\frac{7}{3}, \frac{1}{7}\right)$ and $\left(-\frac{7}{3}, \frac{3}{2}\right)$

  (a) .... **(1 mark)**

  (b) .... **(1 mark)**

  (c) ....

  .... **(2 marks)**

○ **Exam focus**

1   Deduce the equation of the straight line drawn below.

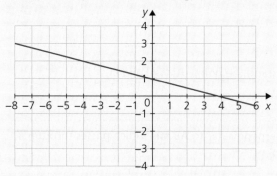

....[3]

2   Calculate the gradient and *y*-intercept for each of the straight lines below:

(a) $3x + y = 5$

....

....[2]

(b) $\dfrac{5x - 3y}{2} = -3$

....

....

....[4]

3   Plot the line given by the equation $x + 6y + 6 = 0$.

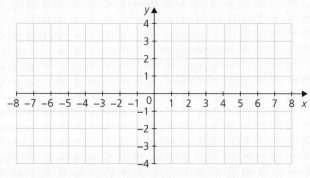

[3]

4   Consider the coordinates of the points A and M:
A (7, 6) and M (10, 2)

(a) Calculate the length of the line segment AM.

....

....[2]

(b) M is the midpoint of the line AB. Calculate the coordinates of the point of B.

....

....[1]

→

5 Calculate the equation of the straight line passing through these two points: (−2, 20) and (2, −4)

....

....

....**[3]**

6 **(a)** Calculate the gradient of the line joining the points $A\left(\frac{1}{2}, 4\right)$ and $B\left(-\frac{3}{2}, 7\right)$.

....

....**[2]**

(b) Calculate the equation of the line perpendicular to AB and which passes through B. Give your answer in the form $ax + by + c = 0$.

....

....

....

....**[4]**

29 Bearings

○ Exercise 29.1

1 A boat sets off from a point A on a bearing of 130° for 4 km to a point B.
At B it changes direction and sails on a bearing of 240° to a point C, 7 km away.
At point C it changes direction again and heads back to point A.

(a) Using a scale of 1 cm : 1 km, draw a scale diagram of the boat's journey.

(4 marks)

(b) From your diagram work out:

(i) the distance AC ....(1 mark)

(ii) the bearing of A from C. .... (2 marks)

(30) Trigonometry

○ Exercises 30.1–30.3

Calculate the value of *x* in each of the diagrams below. Give your answers to 1 d.p.

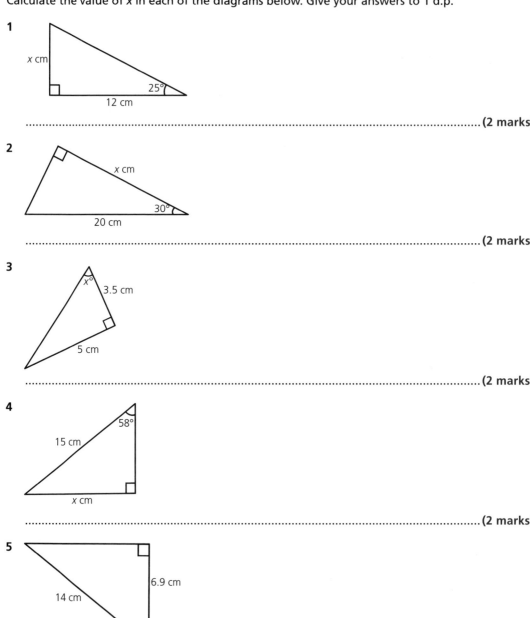

1

x cm

25°

12 cm

....(2 marks)

2

x cm

30°

20 cm

....(2 marks)

3

x°

3.5 cm

5 cm

....(2 marks)

4

58°

15 cm

x cm

....(2 marks)

5

6.9 cm

14 cm

x°

....(2 marks)

6

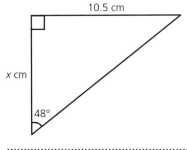

....**(2 marks)**

7

....**(2 marks)**

8

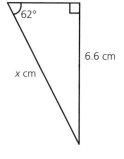

....**(2 marks)**

◯ **Exercises 30.4–30.5**

In each of the following, calculate the length of the marked side, giving your answer to 1 d.p.

1

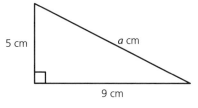

....**(2 marks)**

→

2

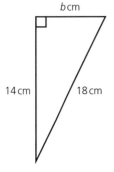

.... **(2 marks)**

3

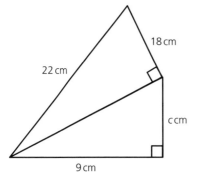

.... **(3 marks)**

4

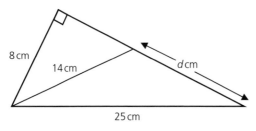

.... **(3 marks)**

5 Three towns, A, B and C, are positioned relative to each other as follows:
- Town B is 68 km from A on a bearing of 225°.
- Town C is on a bearing of 135° from A.
- Town C is on a bearing of 090° from B.

(a) Drawing a sketch if necessary, deduce the distance from A to C.

.... **(2 marks)**

(b) Calculate the distance from B to C.

....

.... **(2 marks)**

6   A boat starts at a point P and heads due north for 20 km to a point Q. At Q it heads east for 15 km to a point R. At R it heads on a bearing of 045° for 10 km to a point S.

   (a) Drawing a sketch if necessary, calculate the **horizontal** distance between R and S.

.... **(2 marks)**

   (b) Calculate the **vertical** distance between P and S.

....

.... **(2 marks)**

   (c) Calculate the shortest distance between P and S.

....

.... **(2 marks)**

   (d) Calculate the bearing of S from P. Give the answer to the nearest degree.

....

.... **(2 marks)**

7   Two trees, A and B, are standing on flat ground 12 m apart as shown. The tops of the two trees are 16 m apart. The angle of elevation of the top of tree A to the ground at X is 50°.

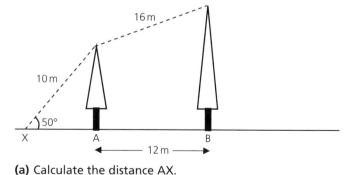

   (a) Calculate the distance AX.

....

.... **(2 marks)**

→

(b) Calculate the height of tree B.

....

....

.... **(3 marks)**

(c) Calculate the angle of depression from the top of tree B to the top of tree A.

....

.... **(2 marks)**

○ Exercise 30.6

1 A point A is at the top of a vertical cliff, 25 m above sea level as shown below. Two points X and Y are in the sea. The angle of elevation from Y to A is 23°. Y is twice as far from the cliff as X.

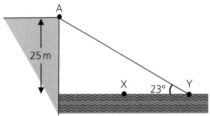

(a) Calculate the horizontal distance of Y from the foot of the cliff.

....

.... **(2 marks)**

(b) Calculate the angle of **depression** from A to X to the nearest whole number.

....

.... **(3 marks)**

(c) Calculate the ratio of the distances AX : AY. Give your answer in the form 1 : *n* where *n* is given to 1 d.p.

....

....

....

....

.... **(5 marks)**

2   A tall vertical mast is supported by two wires, AC and BC, as shown.
    Points A and B are 2.5 m and 6 m above horizontal ground level respectively.
    Horizontally, the mast is 20 m and 27 m from A and B respectively.
    The angle of elevation of C from A is 30°.

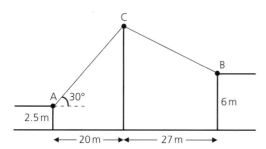

(a) Calculate the height of the mast.

....

....

.... (3 marks)

(b) Calculate the angle of elevation of C from B.

....

....

.... (3 marks)

(c) Calculate the shortest distance between A and B.

....

....

.... (2 marks)

○ **Exercises 30.7–30.8**

1 Express the following in terms of the sine of another angle between 0° and 180°:

   (a) sin 86° .... (1 mark)

   (b) sin 158° .... (1 mark)

2 Express the following in terms of the cosine of another angle between 0° and 180°:

   (a) cos 38° .... (1 mark)

   (b) cos 138° .... (1 mark)

3 Find the two angles between 0° and 180° which have the following sine. Give each answer to the nearest degree.

   (a) 0.37 .... (2 marks)

   (b) 0.85 .... (2 marks)

4 The cosine of which acute angle has the same value as:

   (a) −cos 162° .... (2 marks)

   (b) −cos 136° .... (2 marks)

(31) Further trigonometry

○ **Exercises 31.1–31.2**

1 Calculate the length of the side marked *x*.

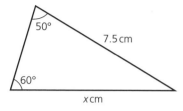

.... **(2 marks)**

2 Calculate the length of the side marked *x*.

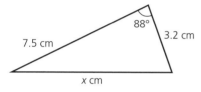

.... **(2 marks)**

3 Calculate the length of the side marked *x*.

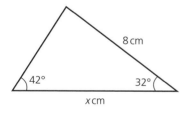

.... **(3 marks)**

4 Calculate the size of the angle marked *θ*.

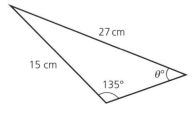

.... **(2 marks)**

5 Calculate the size of the angle θ.

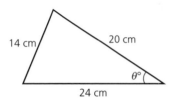

.... **(3 marks)**

6 Calculate the size of the angle θ below, given that it is an obtuse angle (between 90° and 180°).

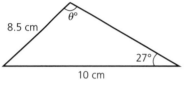

.... **(4 marks)**

○ **Exercise 31.3**

1 A bird, B, flies above horizontal ground.
The bird is 126 m from point A on the ground and 88 m from a point C also on the ground.

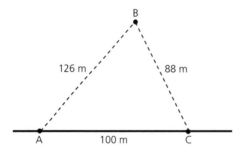

Given that the distance between A and C is 100 m, calculate:

(a) the angle of elevation from A to B

....

....

.... **(3 marks)**

(b) the height of the bird above the ground.

....

....

.... **(2 marks)**

○ **Exercise 31.4**

1 Calculate the area of the triangle below.

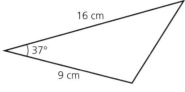

16 cm

37°

9 cm

.... **(2 marks)**

2 A triangle and rectangle are joined as shown below. If the total area of the combined shape is 110 cm², calculate the length of the side marked *x*.

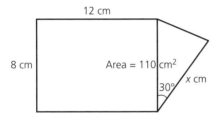

12 cm

8 cm   Area = 110 cm²

30°   *x* cm

.... **(4 marks)**

○ **Exercises 31.5–31.6**

1 The cone below has its apex P directly above the centre of the circular base X.
PQ = 12 cm and angle PQX = 72°.

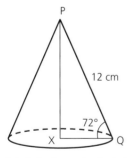

P

12 cm

72°

X   Q

(a) Calculate the height of the cone.

....

.... **(2 marks)**

(b) Calculate the circumference of the base.

....

....

.... **(3 marks)**

→

2 The cuboid ABCDEFGH is shown below.
AD = 8 cm, DH = 5 cm and X is the midpoint of CG.

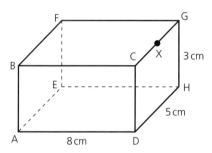

Calculate the following:

(a) the length EG

....

.... **(2 marks)**

(b) the angle EGA

....

.... **(2 marks)**

(c) the length AX

....

.... **(2 marks)**

(d) the angle AXE

....

....

.... **(3 marks)**

3 A right-angled triangular prism ABCDEF is shown below.
AB = 3 cm, AC = 4 cm, BE = 9 cm and point X divides BE in the ratio 1 : 2.

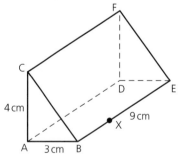

Calculate the following:

(a) the length BC

.... **(1 mark)**

(b) the angle BXC

....

.... **(2 marks)**

(c) the length XF

....

.... **(2 marks)**

(d) the angle between XF and the plane ABDE

....

....

.... **(3 marks)**

4 The diagram below shows a right pyramid, where E is vertically above X.
AB = 5 cm, BC = 4 cm, EX = 7 cm and P is the midpoint of CE.

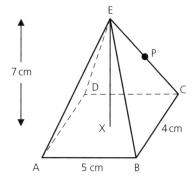

Calculate:

(a) the length AX

....

.... **(2 marks)**

(b) the angle XCE

....

.... **(2 marks)**

(c) the length XP

....

....

....

....

.... **(5 marks)**

○ **Exam focus**

1 If the bearing from X to Y is 127°, calculate the back bearing Y to X.

....

.... [2]

2 Calculate the value of x in the triangle below.

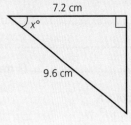

.... [2]

3 Calculate the value of θ to the nearest degree.

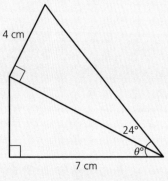

.... [3]

4 Calculate the distance AB.

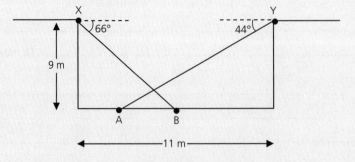

.... [5]

5 θ is an obtuse angle. If sin θ = 0.75, calculate θ to the nearest degree.

....

.... [2]

6 A field PQRS has dimensions and angles as shown.

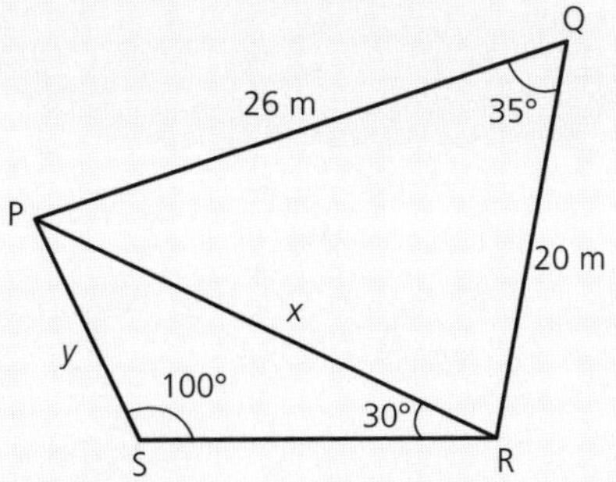

(a) Calculate *x*.

....

....

....**[3]**

(b) Calculate *y*.

....

....**[2]**

(c) Calculate the total area of the field. Give your answer to the nearest whole number.

....

....

....

....**[4]**

7 A right-angled triangular prism is shown below.

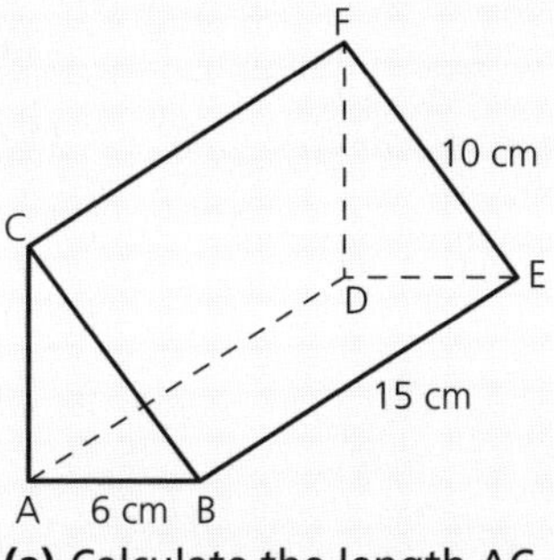

(a) Calculate the length AC.

....

....**[2]**

(b) Calculate the angle CE makes with the base ABDE.

....

....

....

....**[4]**

(32) Vectors

◯ Exercise 32.1

1

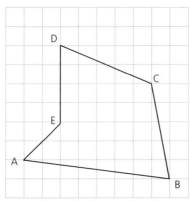

Describe each of the following translations using a column vector:

(a) $\overrightarrow{AB}$ ....(1 mark)

(b) $\overrightarrow{BC}$ ....(1 mark)

(c) $\overrightarrow{CD}$ ....(1 mark)

(d) $\overrightarrow{DE}$ ....(1 mark)

(e) $\overrightarrow{EA}$ ....(1 mark)

◯ Exercises 32.2–32.3

In questions 1 and 2 consider the following vectors:

$$a = \begin{pmatrix} 2 \\ 0 \end{pmatrix} \quad b = \begin{pmatrix} -3 \\ 1 \end{pmatrix} \quad c = \begin{pmatrix} 3 \\ -2 \end{pmatrix}$$

1 Express the following as a single column vector:

(a) $3a$ ....(1 mark)

(b) $2c - b$ .... (2 marks)

(c) $\frac{1}{2}(a - b)$ .... (2 marks)

(d) $-2b$ .... (2 marks)

2 Draw vector diagrams to represent the following:

(a) 2**a** + **b**

(b) −**c** + **b**

(3 marks)

(3 marks)

○ **Exercise 32.4**

1 Consider the vectors:

$$\mathbf{a} = \begin{pmatrix} -2 \\ 0 \end{pmatrix} \quad \mathbf{b} = \begin{pmatrix} -3 \\ 2 \end{pmatrix} \quad \mathbf{c} = \begin{pmatrix} 4 \\ -4 \end{pmatrix}$$

Calculate the magnitude of the following, giving your answers to 1 d.p.

(a) **a** + **b** + **c**

....

....

.... **(3 marks)**

(b) 2**b** − **c**

....

....

.... **(3 marks)**

○ **Exercises 32.5–32.7**

1 Consider the vector diagram below.

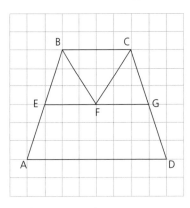

If $\overrightarrow{AB}$ = **a** and $\overrightarrow{BC}$ = **b**, express the following in terms of **a** and **b**.

(a) $\overrightarrow{AE}$ ....(1 mark)

(b) $\overrightarrow{EG}$ ....(1 mark)

(c) $\overrightarrow{AF}$ ....

....(2 marks)

(d) $\overrightarrow{CG}$ ....

....

....(3 marks)

2 In the diagram below, $\overrightarrow{AB}$ = **a**, $\overrightarrow{BD}$ = **b**, D divides the line BC in the ratio 2 : 1 and E is the midpoint of AC.

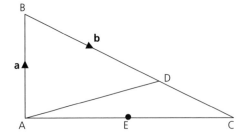

Express the following in terms of **a** and **b**:

(a) $\overrightarrow{AD}$ ....(1 mark)

(b) $\overrightarrow{DC}$ ....(1 mark)

(c) $\overrightarrow{AC}$ ....(2 marks)

(d) $\overrightarrow{ED}$ ....(2 marks)

(33) Matrices

○ Exercise 33.1

1 Give the order of the following matrices:

   (a) $P = \begin{pmatrix} 1 & 3 & 7 & 3 \\ 2 & 0 & 4 & 3 \end{pmatrix}$ ....(1 mark)

   (b) $T = \begin{pmatrix} 2 \\ 1 \\ 3 \\ 0 \end{pmatrix}$ ....(1 mark)

2 The hair colour of students in two classes is recorded. In one class, there are 6 black, 6 brown, 4 blonde and 2 ginger haired students. In the other class there are 8 black, 3 brown and 5 blonde haired students.
   Write this information in a 4×2 matrix.

   (2 marks)

○ Exercise 33.2

Evaluate the following calculations.

1 $\begin{pmatrix} 5 & -1 & 0 \\ 2 & -3 & 4 \end{pmatrix} + \begin{pmatrix} -2 & 0 & 1 \\ -2 & -4 & 6 \end{pmatrix}$ ....(2 marks)

2 $\begin{pmatrix} 7 & -2 \\ 3 & 0 \\ -1 & 4 \end{pmatrix} - \begin{pmatrix} 3 & 2 \\ 3 & -4 \\ -2 & 5 \end{pmatrix}$ ....(2 marks)

3 Michael Phelps, a swimmer from the USA, won 6 gold medals and 2 bronze medals at the 2004 Olympic games in Athens, 8 gold medals at the 2008 Olympic games in Beijing and 4 gold medals and 2 silver medals at the 2012 Olympic games in London.

   (a) Express this information as the sum of three 1×3 matrices.

   .... (2 marks)

   (b) Write down the matrix to represent the total number of each type of medal Phelps won in the three Olympic games.

   .... (1 mark)

○ Exercises 33.3–33.5

1 Evaluate:

(a) $4\begin{pmatrix} 1 & 3 \\ -2 & 0 \\ 2 & \frac{1}{2} \end{pmatrix}$ .... (1 mark)

(b) $-3\begin{pmatrix} 0 & -4 & 2 \\ 1 & -3 & -\frac{1}{3} \end{pmatrix}$ .... (1 mark)

2 Evaluate:

(a) $\begin{pmatrix} -3 & 1 & 6 \end{pmatrix}\begin{pmatrix} -2 \\ 0 \\ 1 \end{pmatrix}$ .... (2 marks)

(b) $\begin{pmatrix} 1 & -2 & 3 \\ -4 & 5 & -6 \end{pmatrix}\begin{pmatrix} 6 & -5 \\ -4 & 3 \\ 2 & -1 \end{pmatrix}$ .... (3 marks)

3 The order of three matrices are given below:
Matrix **A**: (3×2) Matrix **B**: (4×3) Matrix **C**: (2×4)
(a) Give an example of a matrix multiplication that is possible using the matrices above.
.... (1 mark)
(b) Give an example of a matrix multiplication that is not possible using the matrices above.
.... (1 mark)

○ Exercises 33.6–33.7

1 The matrix **A** is $\begin{pmatrix} 0 & 2 \\ 4 & -1 \\ 3 & -1 \end{pmatrix}$.

If $\mathbf{AB} = \begin{pmatrix} 0 & 2 \\ 4 & -1 \\ 3 & -1 \end{pmatrix}$:

(a) Write down matrix **B**.

(1 mark)
(b) What is the name usually given for matrix **B**?
.... (1 mark)

2 Calculate the determinant of the following matrices:

(a) $\begin{pmatrix} 4 & 4 \\ -2 & 1 \end{pmatrix}$ ....

.... (2 marks)

(b) $\begin{pmatrix} 0 & 5 \\ -1 & 6 \end{pmatrix}$ ....

.... (2 marks)

3 If $\mathbf{A} = \begin{pmatrix} 3 & -1 \\ 4 & 0 \\ -2 & 2 \end{pmatrix}$ and $\mathbf{B} = \begin{pmatrix} 1 & 0 & -2 \\ 3 & 6 & -1 \end{pmatrix}$:

(a) calculate **BA**

....

....

.... (3 marks)

(b) evaluate |3**BA**|.

....

....

.... (3 marks)

○ **Exercise 33.8**

1 Using simultaneous equations and giving each element as a fraction, calculate the inverse of the matrix $\begin{pmatrix} 4 & -2 \\ 4 & 2 \end{pmatrix}$.

(4 marks)

2 **(a)** If $\mathbf{A}^{-1} = \begin{pmatrix} 2 & 0 \\ 1 & 2 \end{pmatrix}$, calculate matrix **A**.

(4 marks)

(b) Deduce the matrix produced by $\mathbf{AA}^{-1}$.

(2 marks)

Transformations

◯ Exercises 34.1–34.2

1   On the following diagrams:
   (i)   draw the position of the mirror line(s)
   (ii) give the equation of the mirror line(s).

(a) (i)

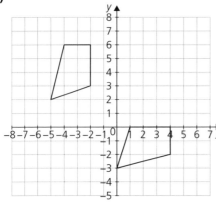

(b) (i)

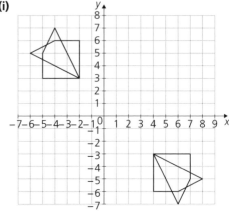

(ii)...(2 marks)   (ii)...(4 marks)

2   Reflect the following object in the line $y = -x + 1$.

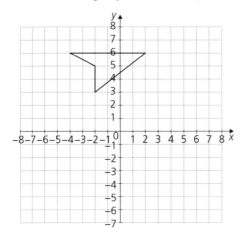

(2 marks)

○ Exercises 34.3–34.4

1 In the following, the object and centre of rotation have been given. Draw the object's image under the stated rotation.

(a)

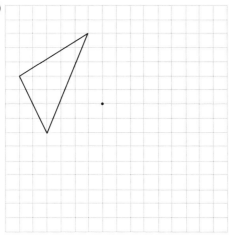

Rotation 180°   **(2 marks)**

(b)

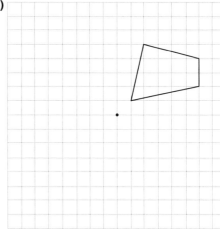

Rotation 90° anticlockwise   **(2 marks)**

2 In the following, the object (unshaded) and image (shaded) have been drawn. In each diagram, mark the centre of rotation and calculate the angle and direction of rotation.

(a)

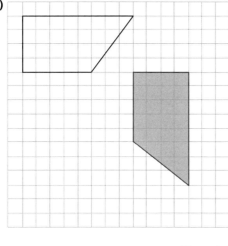

.... (2 marks)

(b)

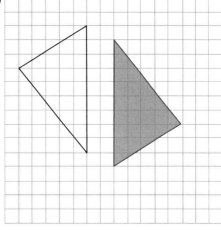

.... (2 marks)

○ **Exercise 34.5**

In the diagrams below, assume the shaded shape is the object. By construction:

    (a) find the centre of rotation and give its coordinates   **(3 marks)**
    (b) find the angle and direction of the rotation.   **(2 marks)**

1

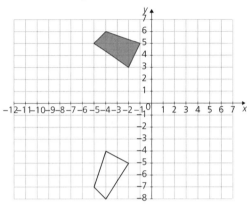

2

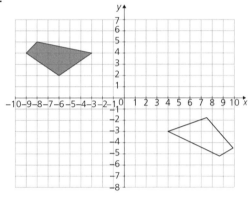

....

○ **Exercises 34.6–34.7**

1 In the following diagram, object A has been translated to each of the images B, C and D. Give the translation vector in each case.

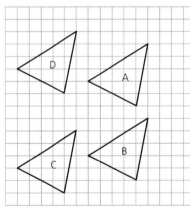

B = ... **(1 mark)**

C = ... **(1 mark)**

D = ... **(1 mark)**

2 In the diagram below, translate the object by the stated vector.

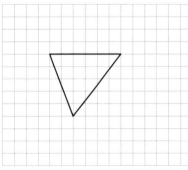

$$\begin{pmatrix} -2 \\ -4 \end{pmatrix}$$

(2 marks)

○ **Exercises 34.8–34.9**

1 Find the centre of enlargement and the scale factor of enlargement in the diagrams below.

(a)

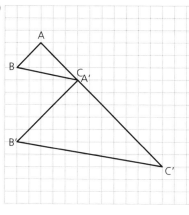

(b)

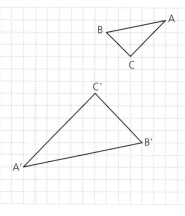

.... (2 marks)   .... (2 marks)

2 Enlarge the object below, by the scale factor given and from the centre of enlargement shown.

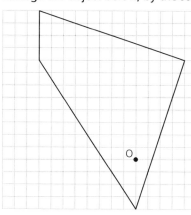

Scale factor $\frac{1}{4}$

(2 marks)

○ **Exercise 34.10**

1 An object and part of its image under enlargement are given in the diagram below.

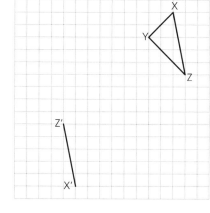

(a) Complete the image.   (2 marks)

(b) Find the centre of enlargement.   (1 mark)

(c) Calculate the scale factor of enlargement.

.... (1 mark)

◯ **Exercise 34.11**

1 In this question draw each transformation on the same grid and label the images clearly.

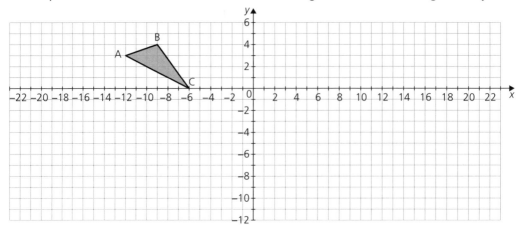

(a) Map the triangle ABC onto $A^1B^1C^1$ by an enlargement of scale factor -2, with the centre of enlargement at $(-4, 2)$. **(3 marks)**

(b) Map the triangle $A^1B^1C^1$ onto $A^2B^2C^2$ by a reflection in the line $x = 4$. **(2 marks)**

(c) Map the triangle $A^2B^2C^2$ onto $A^3B^3C^3$ by a rotation of 180°, with the centre of rotation at $(-4, 2)$. **(2 marks)**

◯ **Exercises 34.12–34.13**

1 The object ABC undergoes a transformation. The transformation matrix is $\begin{pmatrix} -1 & 0 \\ 0 & -1 \end{pmatrix}$.

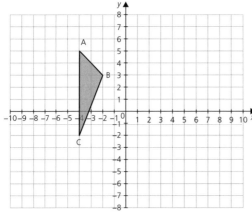

(a) Express the vertices of the triangle ABC as a matrix.

(1 mark)

(b) By carrying out the appropriate calculation, determine in matrix form, the vertices of the image A'B'C'.

(3 marks)

(c) Draw the image A'B'C' on the grid opposite and label each vertex. (1 mark)

(d) (i) Describe in geometrical terms the transformation that maps triangle ABC on to A'B'C'.

....(1 mark)

(ii) Describe in geometrical terms a different transformation that maps triangle ABC on to A'B'C'.

....(1 mark)

2   The grid below shows a triangle ABC.

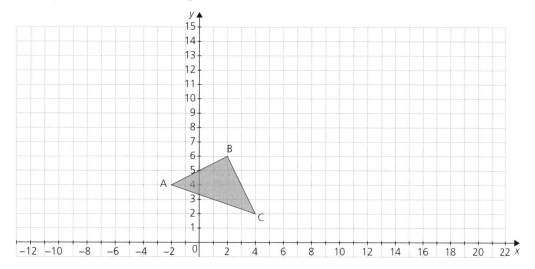

(a) Calculate the area of triangle ABC.

.... (2 marks)

(b) Draw the image of ABC under the transformation of the matrix $\begin{pmatrix} -\frac{5}{2} & 0 \\ -1 & -\frac{5}{2} \end{pmatrix}$ and label the vertices A'B'C'. (4 marks)

(c) Calculate the determinant of the transformation matrix.

.... (2 marks)

(d) Using your answer to part (c) deduce the area of the image A'B'C'.

.... (1 mark)

(e) What matrix maps A'B'C' on to ABC?

(4 marks)

○ **Exercise 34.14**

1 Two transformation matrices **T** and **U** are stated as follows:

$$\mathbf{T} = \begin{pmatrix} 0 & 1 \\ -1 & 0 \end{pmatrix} \qquad \mathbf{U} = \begin{pmatrix} 0 & -1 \\ -1 & 0 \end{pmatrix}$$

The grid below shows a rectangle ABCD.

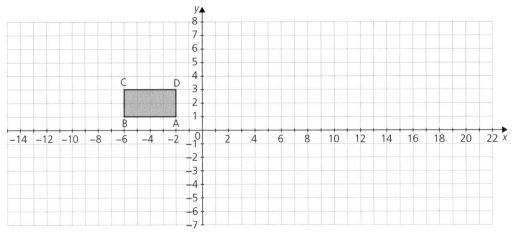

(a) On the grid, draw an image of ABCD under the transformation of matrix **T** and label its vertices A'B'C'D'. **(4 marks)**

(b) On the grid, draw an image of A'B'C'D' under the transformation of matrix **U** and label its vertices A''B''C''D''. **(4 marks)**

(c) Determine the single transformation matrix that would map ABCD on to A''B''C''D''.

(3 marks)

(d) Determine the single transformation matrix that would map A''B''C''D'' on to ABCD.

(4 marks)

○ **Exam focus**

1   If $\mathbf{a} = \begin{pmatrix} 5 \\ -2 \end{pmatrix}$ and $\mathbf{b} = \begin{pmatrix} -1 \\ 4 \end{pmatrix}$, write the column vector for:

(a) $2\mathbf{a} - \mathbf{b}$ ....[2]

(b) $\frac{1}{2}(\mathbf{b} - \mathbf{a})$ ....[2]

2   The position vector of P is $\begin{pmatrix} 3 \\ 1 \end{pmatrix}$, the position vector of Q is $\begin{pmatrix} -2 \\ -3 \end{pmatrix}$.

(a) Calculate the vector $\overrightarrow{PQ}$. ....[2]

(b) Calculate $\left| \overrightarrow{PQ} \right|$. ....

....[3]

3   The rhombus ABCD is shown below.

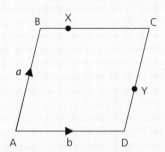

$\overrightarrow{AB} = \mathbf{a}$ and $\overrightarrow{AD} = \mathbf{b}$. X divides the line BC in the ratio 1:3 and Y divides the line CD in the ratio 2:1. Express the following in terms of $\mathbf{a}$ and $\mathbf{b}$:

(a) $\overrightarrow{BC}$ ....[1]

(b) $\overrightarrow{AX}$ ....[2]

(c) $\overrightarrow{AY}$ ....[2]

(d) $\overrightarrow{YX}$ ....[3]

→

4 The graph below shows the rainfall (mm) per quarter over the six year period 2007–2012.

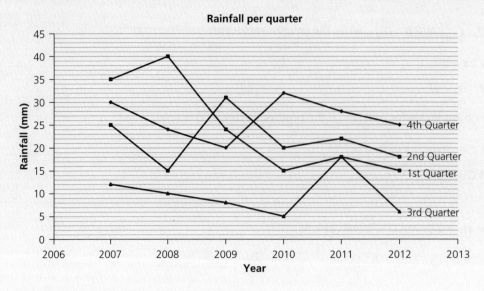

Represent the data in a 6 × 4 matrix.

.... **[3]**

5 Two matrices **P** and **Q** are given as follows:

$$\mathbf{P} = \begin{pmatrix} 3 & 6 & -1 \\ 4 & 0 & -2 \end{pmatrix} \text{ and } \mathbf{Q} = \begin{pmatrix} -1 & 0 \\ 2 & 0 \\ 3 & -2 \\ 0 & -4 \end{pmatrix}$$

(a) Which product of the two matrices is possible? Justify your answer.

....

....**[2]**

(b) Carry out the multiplication stated in part (a) above.

....**[3]**

6 $\mathbf{A} = \begin{pmatrix} 2 & -2 \\ 0 & 1 \end{pmatrix}$ and $\mathbf{B} = \begin{pmatrix} -1 & 0 \\ 2 & 0 \end{pmatrix}$

(a) Evaluate $|\mathbf{A} - \mathbf{B}|$.

....[3]

(b) Calculate the matrix $(\mathbf{A} - \mathbf{B})^{-1}$.

....[4]

7 In the diagram below, the object ABCD is mapped on to A'B'C'D'.

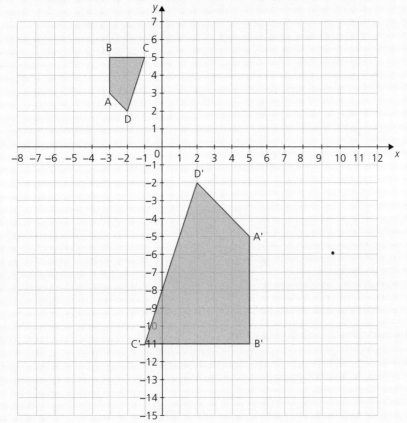

(a) By construction, find the position and coordinates of the centre of enlargement. Mark it on the diagram and label it O. [2]

(b) Calculate the scale factor of enlargement.

....[1]

(c) If the area of the object is 8 units², and using your answer from (b), deduce the area of the image A'B'C'D'.

....[2]

8 In the diagram below, the shaded shape is the object and it is mapped on to its image by rotation.

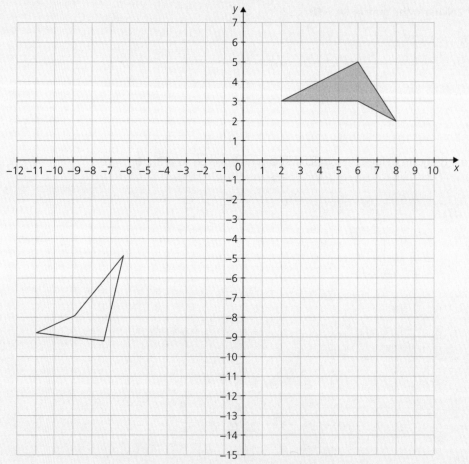

(a) By construction, find the coordinates of the centre of rotation. ....[3]

(b) Find the angle and direction of the rotation. ....

....[2]

9 The object ABC is mapped on to A'B'C' by the transformation matrix $\begin{pmatrix} 2 & 0 \\ 0 & 2 \end{pmatrix}$.

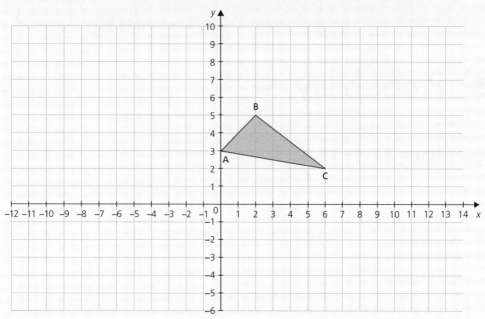

(a) (i) Determine in matrix form the vertices of the image A'B'C'.

....[3]

   (ii) Plot the image A'B'C' on the grid above, labelling the vertices clearly.   [1]

(b) A'B'C' is mapped onto A''B''C'' by the transformation matrix $\begin{pmatrix} 0 & -1 \\ 1 & 0 \end{pmatrix}$.
   Plot the image A''B''C'' on the same grid, labelling the vertices clearly.   [3]

(c) Determine the single matrix that will map ABC on to A''B''C''.

....[2]

(d) If the area of ABC is x units^2, determine the area of A''B''C'' in terms of x.

....

....[2]

(e) Calculate the single matrix that will map A''B''C'' on to ABC.

....[4]

Probability

(35) Probability

○ **Exercises 35.1–35.4**

1 Calculate the theoretical probability, when rolling an octahedral, fair dice, of getting each of the following:

(a) a score of 6...(1 mark)

(b) a score of 2 or 3 ....(1 mark)

(c) an even number ....(1 mark)

(d) a score less than 1 ....(1 mark)

(e) a score of more than 1 ....(1 mark)

(f) a score less than 4 or more than 4 ....(1 mark)

2 (a) Calculate the probability of being born in June. (Assume a year is 365 days.)

....(1 mark)

(b) Explain why the answer to part (a) is not $\frac{1}{12}$.

....(1 mark)

(c) What is the probability of not being born in June?.................................(1 mark)

3 A container has 749 white sweets and 1 red sweet. What is the probability of picking the red sweet if a child randomly picks:

(a) 1 sweet ....(1 mark)

(b) 500 sweets ....(1 mark)

(c) 150 sweets ....(1 mark)

(d) 750 sweets? ....(1 mark)

4 In a class there are 23 girls and 17 boys. They enter the room in a random order. Calculate the probability that the first student to enter will be:

(a) a girl ....(1 mark)

(b) a boy. ....(1 mark)

5 Tiles, each lettered with one different letter of the word MATHEMATICS, are put into a bag. If one tile is drawn out at random, calculate the probability that it is:

(a) an A or M or T ....(1 mark)

(b) not a consonant ....(1 mark)

(c) not an X, Y or Z ....(1 mark)

(d) not a letter in your first name. ....(1 mark)

6 **(a)** Three red, 17 white, 25 blue and 7 green counters are put into a bag. If one is picked at random, calculate the probability that it is:

  (i) a green counter ....(1 mark)

  (ii) not a blue counter. ....(1 mark)

 (b) If the first counter taken out is green and it is not put back into the bag, calculate the probability that the second counter picked is:

  (i) not a green counter ....(1 mark)

  (ii) a red counter. ....(1 mark)

7 The letters of the word MATHEMATICS are written on individual cards. A card is chosen at random, it is then replaced and a second card is chosen. What is the probability of choosing:

(a) an M twice ....

.... (3 marks)

(b) an S followed by a T?..

.... (3 marks)

8 A computer uses the letters A, T or R at random to make three-letter words. Assuming that a letter can be repeated, calculate the probability of getting:

(a) the letters R,R,R ....

.... (3 marks)

(b) any one of the words TAR, RAT or ART.

....

.... (3 marks)

→

9 The gender and age of members of a film club are recorded and shown below.

	Child	Adult	Senior
Male	14	58	21
Female	18	44	45

(a) How many members has the film club got?

.... (1 mark)

(b) A member is picked at random. Calculate the probability that it is:

(i) a child ....(1 mark)

(ii) female ....(1 mark)

(iii) a female child. ....(1 mark)

10 Two friends are standing in a hall with many other people. A person is picked randomly from the hall.
How many people are in the hall if the probability of either of the friends being picked is 0.008?

....

....(2 marks)

(36) Further probability

○ Exercise 36.1

1 A fair cubic dice and a fair tetrahedral dice are rolled.
Use a two-way table if necessary to find:

(a) the probability that both dice show the same number

.... **(2 marks)**

(b) the probability that the number on one dice is double the number on the other.

.... **(2 marks)**

2 Two fair octahedral dice are rolled.
Use a table if necessary to find the probability of getting:

(a) any double

.... **(2 marks)**

(b) a total score of 13 .... **(2 marks)**

(c) a total score of 17 .... **(2 marks)**

(d) a total which is either a multiple of 2, 3 or 5. ....

.... **(4 marks)**

○ **Exercise 36.2**

1   A football team plays three matches. In each match the team is equally likely to win, lose or draw.

   (a) Calculate the probability that the team:

   (i) wins no matches ....

   .... **(3 marks)**

   (ii) loses at least two matches. ....

   .... **(3 marks)**

   (b) Explain why it is not very realistic to assume that the outcomes are equally likely in this case of football matches.

   .... **(1 mark)**

2   A spinner is split into fifths, numbered 1–5.
   If it is spun twice, calculate the probability of getting:

   (a) two fives .... **(2 marks)**

   (b) two numbers the same ....

   .... **(3 marks)**

○ **Exercise 36.3**

1   A particular board game involves players rolling an octahedral dice. However, before a player can start, he or she needs to roll an odd number.
   Draw a tree diagram if necessary (there is blank space for this on the page opposite) to calculate the probability of the following:

   (a) getting an eight on the first throw .... **(2 marks)**

   (b) starting within the first two throws

   .... **(3 marks)**

   (c) not starting within the first three throws

   .... **(3 marks)**

   (d) starting within the first three throws.

   .... **(3 marks)**

   (e) If you add the answers to (c) and (d) what do you notice? Explain this result.

   ....

   .... **(2 marks)**

2   In England 40% of trucks are foreign made. Calculate the following probabilities:

   (a) the next two trucks to pass a particular spot are both English

   .... (3 marks)

   (b) two of the next three trucks are foreign

   .... (3 marks)

   (c) two or more of the next three trucks are English.

   .... (3 marks)

3   The first team of Barcelona F.C. has a 0.05 chance of losing a game.
    Calculate the probability of the team achieving:

   (a) two consecutive losses

   .... (2 marks)

   (b) 20 consecutive wins.

   .... (3 marks)

○ **Exam focus**

1   A dice is a regular dodecahedron. It is numbered 1–12.

(a) What is the probability of throwing a 6 with the first throw?

....[1]

(b) What is the probability of throwing a prime number? (1 is not a prime)

....[1]

(c) What is the probability of throwing two 1's in the first two throws?

....[2]

(d) What is the probability of not throwing two 1's in the first two throws?

....[1]

2   An icosahedral dice is numbered 1 to 20. It is thrown and a fair coin is spun.
Find the probability:

(a) of getting a head and a factor of 20

....

....[2]

(b) of getting a prime number and a tail.

....

....[2]

3   A computer uses the letters A, E or T at random to make three-letter words.
Assuming that a letter can be repeated, calculate the probability of getting:

(a) the letters AAA

....[2]

(b) the word EAT, TEA, ATE or TEE.

....

....[3]

4   A top tennis player calculates he has a 20% chance of losing a set of tennis. As a percentage:

(a) What is the probability that he loses 3 consecutive sets?

....

....[3]

(b) What is the probability that he wins three consecutive sets?

....

....[3]

5   A bag contains *r* red balls and *b* blue balls. Two balls are chosen.
If the first ball is not replaced after being picked, what is the probability of choosing:

(a) a red ball first

....[3]

(b) a red ball first and a red ball second

....

....[4]

(c) one ball of each colour?

....

....[4]

(37) Mean, median, mode and range

○ Exercises 37.1–37.2

1 A student looks at the results of her last ten maths tests. Each score is out of 10.

   6 4 9 8 8 3 4 5 8 6

   Calculate:

   (a) the mean test score

   ....(1 mark)

   (b) the median test score

   ....(1 mark)

   (c) the modal test score ....(1 mark)

   (d) the test score range ....(1 mark)

2 The mean mass of 15 rugby players in a team is 115.3 kg.
   The mean mass of the team plus a substitute is 114 kg.
   Calculate the mass of the substitute.

   ....

   ....(2 marks)

3 A chocolate manufacturer sells, as one of its products, a box with assorted chocolates.
   A number of these boxes are sampled and the number of chocolates inside recorded.
   The results are shown below.

Number of chocolates	42	43	44	45	46	47	48
Frequency	3	7	8	7	9	5	1

   (a) How many boxes were sampled?

   ....(1 mark)

   (b) What is the modal number of chocolates? ....(1 mark)

(c) Calculate the mean number of chocolates.

....

.... (2 marks)

(d) Calculate the median number of chocolates. .... (1 mark)

(e) Calculate the range of the number of chocolates. .... (1 mark)

◯ Exercise 37.3

1 A school holds a sports day. The time taken for a group of students to finish the 1 km race is shown in the grouped frequency table below.

Time/mins	4–	5–	6–	7–	8–9
Frequency	1	4	8	7	2

(a) How many students completed the race? .... (1 mark)

(b) Estimate the mean time it took for the students to complete the race.
Give your answer to the nearest second.

....

....

....

.... (4 marks)

(38) Collecting and displaying data

○ Exercises 38.1–38.3

1 In 2012, the Olympics were held in London. 15 athletes were chosen at random and their height (cm) and mass (kg) were recorded. The results are shown below.

Height/cm	Mass/kg
201	120
203	93
191	97
163	50
166	63
183	90
182	76
183	87

Height/cm	Mass/kg
166	65
160	41
189	82
198	106
204	142
179	88
154	53

(a) What type of correlation (if any) would you expect between a person's height and mass? Justify your answer.

....

.... (2 marks)

(b) Plot a scatter graph on the grid below.

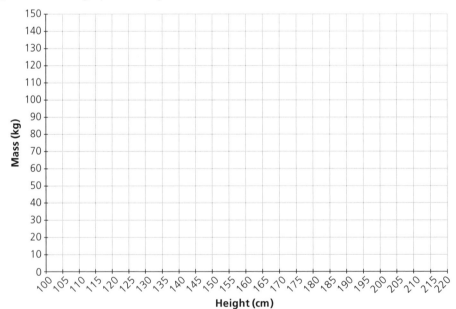

Height (cm)

(3 marks)

(c) (i) Calculate the mean height of the athletes.

....(1 mark)

(ii) Calculate the mean mass of the athletes.

....(1 mark)

(iii) Plot the point representing the mean height and mean mass of the athletes. Label it M. (1 mark)

(d) Draw a line of best fit for the data, making sure it passes through the mean point M.

(1 mark)

(e) (i) From the results you have plotted, describe the correlation between the height and mass of the athletes.

....(1 mark)

(ii) How does the correlation compare with your prediction in (a)?

....(1 mark)

○ Exercises 38.4–38.5

1 The ages of 80 people, selected randomly, travelling on an aeroplane are given in the grouped frequency table below.

Age (years)	0–	15–	25–	35–	40–	50–	60–	80–100
Frequency	10	10	10	10	10	10	10	10
Frequency density								

(a) Complete the table above by calculating the frequency density. **(2 marks)**

(b) Represent the information as a histogram on the grid below.

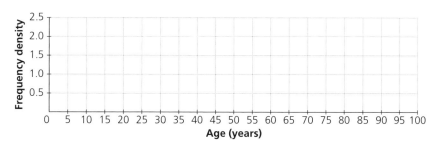

(3 marks)

Cumulative frequency

○ Exercises 39.1–39.2

1 A candle manufacturer wishes to test the consistency of his candles by seeing how long they last. He randomly selects 160 candles, lights them and records how long they last in minutes. The results are presented in the grouped frequency table below.

Time (mins)	140–	150–	160–	170–	180–	190–	200–	210–220
Frequency	5	20	45	30	25	20	10	5
Cumulative frequency								

(a) Complete the table above by calculating the cumulative frequency.　　　　　**(1 mark)**

(b) Plot a cumulative frequency graph on the axes below.

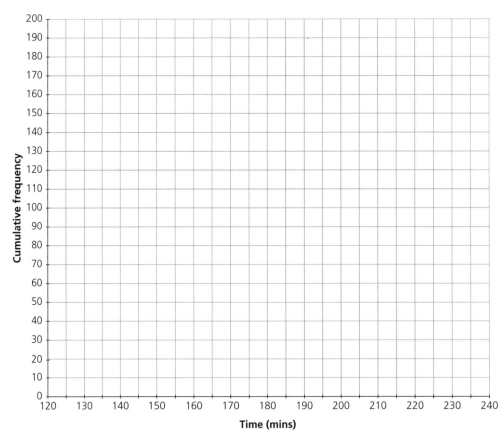

(3 marks)

(c) From your graph, estimate the median amount of time that the candles last.

.... (1 mark)

(d) From your graph estimate:

   (i) the upper quartile time .... (1 mark)

   (ii) the lower quartile time .... (1 mark)

   (iii) the interquartile range. .... (1 mark)

(e) The candle manufacturer is aiming that the lifespans of the middle 50% of his candles do not differ by more than 30 minutes. Explain, giving your justification, whether the data supports his aim.

....

.... (2 marks)

○ **Exam focus**

1 Eight people are weighed. Their masses (kg) are given below.

75 *x* 92 46 71 84 *y* 97

Two of their masses are unknown and are given as *x* and *y*.

The mean of the 8 masses is known to be 74 kg, the median 73 kg and the range 51 kg. Calculate a possible pair of values for *x* and *y*.

....

....

.... [3]

2 Eight different cars are selected at random. Their mass (kg) is recorded, as is their average fuel consumption (kpl). The results are presented in the table below.

Mass (kg)	2150	1080	1390	1820	900	1210	1620	810
Average fuel consumption (kpl)	5.1	11.2	7.8	6.5	15.1	10.7	11.6	12.0

(a) Plot a scatter graph on the grid below.

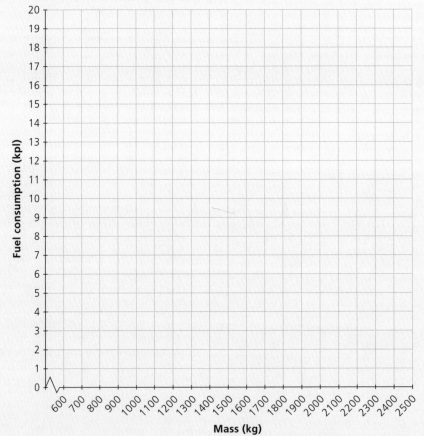

Mass (kg)   [3]

(b) Describe the correlation between mass and fuel consumption

.... [2]

(c) Calculate:

   (i) the mean mass of the sampled cars

   ....[1]

   (ii) the mean fuel consumption of the sampled cars.

   ....[1]

(d) Plot the point that represents the mean mass and mean fuel consumption on your graph and label it M.   [1]

(e) Draw a line of best fit on your graph that passes through M.   [1]

(f) A car manufacturer brings out a new car that weighs only 700 kg. Using your graph, estimate what its average fuel consumption is likely to be.

   ....[2]

3 A bus company wishes to investigate how late its buses on a particular route are. The lateness (mins) of buses over a day is recorded and presented in the grouped frequency table below.

Bus lateness (mins)	0–	0.5–	1–	2–	5–	10–	20–30
Frequency	10	8	16	15	20	10	5
Frequency density							

(a) Complete the table by calculating the frequency density.   [2]

(b) Plot a histogram on the grid below.

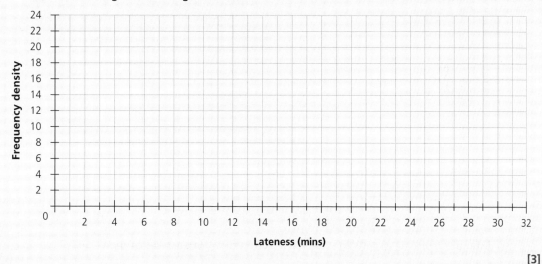

Lateness (mins)

[3]

4 An experiment is set up to investigate the effect of tiredness on the reaction time(s) of adults.

The same test is carried out on each adult, once when they have had a full night's sleep and once when they have not slept.
The results are shown in the tables below.

Full sleep

Reaction time (s)	0–	0.1–	0.2–	0.3–	0.4–	0.5–0.6
Frequency	1	14	18	4	2	1
Cumulative frequency						

No sleep

Reaction time (s)	0–	0.1–	0.2–	0.3–	0.4–	0.5–	0.6–	0.7–0.8
Frequency	0	2	5	15	10	5	2	1
Cumulative frequency								

(a) Complete the cumulative frequency row in each of the tables above. **[2]**

(b) On the grid below, plot cumulative frequency curves for both sets of data. Label each one clearly.

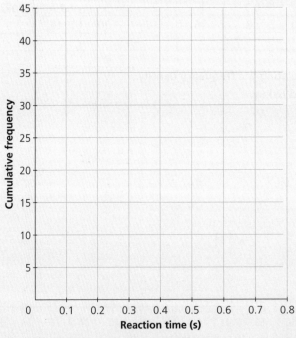

(c) Estimate from the graphs the median reaction time for each set of data.

....**[2]**

(d) Estimate from the graphs the interquartile range for each set of data.

....

....**[4]**

(e) State two conclusions you can make from the results of your calculations in (c) and (d) above.

....

....**[2]**